Critical Reflections on Nuclear and Renewable Energy

Scrivener Publishing
100 Cummings Center, Suite 541J
Beverly, MA 01915-6106

Publishers at Scrivener
Martin Scrivener (martin@scrivenerpublishing.com)
Phillip Carmical (pcarmical@scrivenerpublishing.com)

Critical Reflections on Nuclear and Renewable Energy

Environmental Protection and Safety in the Wake of the Fukushima Nuclear Accident

Way Kuo

President of City University of Hong Kong

Scrivener
Publishing

WILEY

Co-published by John Wiley & Sons, Inc. Hoboken, New Jersey, and Scrivener Publishing LLC, Salem, Massachusetts.
Published simultaneously in Canada.

For general information on our other products and services or for technical support, please contact our Customer Care Department within the United States at (800) 762-2974, outside the United States at (317) 572-3993 or fax (317) 572-4002.

Wiley also publishes its books in a variety of electronic formats. Some content that appears in print may not be available in electronic formats. For more information about Wiley products, visit our web site at www.wiley.com.

For more information about Scrivener products please visit www.scrivenerpublishing.com.

Cover design by Russell Richardson

Library of Congress Cataloging-in-Publication Data:

ISBN 978-1-118-77342-0

Printed in the United States of America

10 9 8 7 6 5 4 3

Contents

Bureaucrats are those who feel disinclined to say what should be said or to do what should be done.
Engineers are those who do not dare to say what should be said, but persist stubbornly in doing what should be done.
Scholars are those who say whatever comes to their mind, but shy away from doing what should be done.
Politicians are people who talk rot without thinking and act rashly and randomly without regard to whether such action should be taken.

Foreword 1

by Arden Bement

President Way Kuo applies his world-recognized expertise in system reliability in sharing his reflections of why energy sustainability is critical for society's well being in the information age. This especially applies to developing countries, which are actively building economic capacity to meet their societal needs. The availability of affordable, reliable, and safe energy will control their rate of progress. However, Way Kuo in a most scholarly way explains that one can't have energy sustainability without concurrently assuring environmental and economic sustainability. Achieving all three will require not only continued innovation but also an improved understanding of the interrelationships among the technical, social (including behavioral) and economic factors involved in building greater energy capacity and distribution on 'smart grids'.

The author approaches his reflections on nuclear and renewable energy as a true scholar, giving the reader extensive evidence for evaluating the viable alternative forms of energy supply for themselves while maintaining a light touch, embracing oriental culture, history and poetry to illustrate his points, when arguing their relative advantages and disadvantages.

Way Kuo in posing his arguments lets facts speak for themselves. However, in terms of net present value of environmental protection, affordability, and safety he concludes that nuclear power offers the best form of energy supply for the future. He devotes an extensive part of his book making this case, informing the reader of the relative impacts and deaths caused by the disastrous use of coal over time for large-scale energy supply. He gives evidence that even so-called 'clean' forms of renewable energy have environmental impacts. He explains in comparative detail the consequences of the major three nuclear disasters over the past forty-five years — at the Three Mile Island Nuclear Reactor in 1979, at the Chernobyl Reactor site in the Ukraine in 1986, and at the Fukushima Daiichi Nuclear Site on March 11, 2011.

In the case of the Three Mile Island and Fukushima disasters, which were designed with full concrete containment meeting international standards, the reader may be surprised to learn — contrary to a widely held view — that not a single life was lost to date. While they certainly may rank as economic disasters, they should not be accounted as disasters that entailed enormous losses of life. In the case of Fukushima Daiichi, less- than-transparent owners and government leaders exacerbated public hysteria shortly after the disaster occurred. In the case of the Chernobyl Plant the explosion that penetrated the reactor building resulted in a human disaster of monumental proportion, with thirty-one deaths among operators and emergency workers and over 4,000 civilian deaths caused by excessive exposure to radiation. Adequate engineered safeguards, such as a containment building meeting international standards, were not provided for this plant. As Way Kuo points out, all three disasters can be attributed to human error, where the operators proved to be the least reliable link in the chain of control for returning the reactors to safe control and shutdown after a major upset.

The author devotes much of his book to discussing human fear of radiation as a key factor in the public's willingness to accept nuclear energy. By pointing out that the thousands of deaths that occur each year in the mining, transportation, and combustion of coal receive little attention he argues that humans are more comfortable with the forms of death that they do understand than those they don't understand, such as exposure to radiation. However, he points out exposure to radiation is part of living. It is ubiquitous and can't be avoided if one wants to live in cities, fly airplanes, eat certain foods, or undergo medical radiation diagnosis and therapy. People actually expose each other by small amounts of radioactivity in their bones and organs.

Finally, the author points out from his experience that the lessons learned from disasters such as Three Mile Island, Chernobyl and Fukushima also have positive outcomes in compelling corrective actions to achieve greater safety and reliability. The design, construction and operation of future nuclear power plants will benefit from the lessons learned from these events.

As a result of Fukushima, regulators are taking firmer measures to routinely inspect nuclear power plants and assess their safety and reliability as a function of aging. They will also be more stringent in testing the effectiveness of training plant operators to respond to upset conditions no matter how rare. Finally, they will recognize that records are made to be broken. There is nothing sacrosanct about a hundred-year accident. Fukushima proved that designing to a hundred-year event was not sufficient. He also points out that even though absolute safety is an unreasonable expectation, that 100% reliable is a time-dependent challenge that must engage well-trained regulators, management, and operators alike. Safety, reliability, and transparency factor in his formula for building and operating nuclear power plants now and in the future.

Arden Bement
David A. Ross Distinguished Professor, Purdue University
Director of National Science Foundation, 2004-2010
Director of National Institute of Standards and Technology, 2001-2004

December 1, 2013

Foreword 2[*]

by Xu Kuangdi

The exploitation of energy resources has been indispensable to human evolution. Though I am not an anthropologist, I believe learning to make use of fire was a defining factor in distinguishing human beings from other animals. Humans, with their ability to use fire, evolved into the most dominant living creatures on earth, and developed a dependence on fossil fuel, especially since the era of the steam and internal combustion engines, for producing food, clothing, shelter and transport.

But in the course of making our lives more prosperous, we have produced huge amounts of greenhouse gases. But burning fossil fuels is threatening the environment we live in. The emission of PM2.5 has reached serious levels, and SO_2, NOx and particles produced during the consumption of fossil energy account for 94%, 60% and 70% respectively of air pollutants. This is why we must strictly control the consumption of fossil fuels such as coal, gas and oil on the one hand, and vigorously develop cleaner energies on the other.

Typical examples of cleaner energies include hydropower, wind-generated power, solar energy, geothermal energy and safe and reliable nuclear energy. Hydropower, which is restricted by water resources and geological conditions, usually accounts for 10 to 20% of the total energy consumption of a country, except in a small number of countries such as Norway and Switzerland which enjoy ample levels of rainfall and snowfall, as well as high mountains and great gorges. Limited in scope, geothermal energy is unlikely to become a main source of energy, except in Iceland, which has an extraordinary geological structure. Wind and solar energy, which are favorite sources of clean energy for the 21st century, generally account for less than 10% of the power capacity of a country. What's more, the

---

[*] The original was written in Chinese.

xiii

production of solar panels and wind turbine blades is expensive, their operation not stable, and their access to the national grid restricted. As a result, these two energies are mostly available in scattered residential areas for the purpose of separate power supplies.

A series of scientific discoveries concerning the structure of the atomic nucleus and nuclear energy at the beginning of the 20th century brought about a historic change not only in the field of physics, but also in the whole scientific and technological world. As a result, people began to look at the microscopic material world from an entirely new perspective and discover the amazing potential of nuclear energy. Against the special historical background of World War Two, the technology of nuclear fission was adopted for the development of nuclear weapons.

The large-scale application of nuclear energy serves as a successful example of the transformation of basic physics into engineering applications. More than 400 nuclear reactors have been built in 20-plus countries. People have accumulated the operational experience of 14,000 reactor-years, and nuclear power accounts for 14% of the world's power supply. Like all other energy industries, the nuclear power industry has witnessed industrial accidents in the course of its development. Calamitous incidents in Three Mile Island, USA, in Chernobyl, former USSR, and Fukushima, Japan, have raised doubts about the safety of nuclear power and its environmental consequences, greatly dampening enthusiasm for the development of nuclear power.

Professor Way Kuo, President of City University of Hong Kong, is a famous expert in the field of reliability. Using the Fukushima nuclear incident as a point of penetration, Professor Kuo has adopted the method of reliability analysis for engineering safety in nuclear energy and environmental protection. He has succeeded in analysing the abstruse principle of this hi-tech problem, about which opinions differ greatly. The plain and vivid terms he uses in his writing will make his readers easily understand the complex ideas involved. I am sure his humorous style will evoke smiles from his readers, too, which reminds me of a remark made by the late Mr Qian Xuesen: "An expert is one who can explain precisely the problems about special science, but his audience may find it difficult to comprehend his explanation. A great master is one who can explain precisely the problems about special science, and his audience will find it easy to comprehend his explanation."

In my opinion, Professor Way Kuo's book has reached a very high standard of science. What's more, it distinguishes itself by its readability and comprehensibility. Professor Kuo's book is so reader-friendly that it can help dispel the doubts of its readers.

Xu Kuangdi
Honorary Chairman of the Chinese Academy of Engineering
President of the Chinese Academy of Engineering, 2002–2010
Mayor of Shanghai, 1995–2001

January 1, 2014

Preface

Nuclear energy has been a controversial topic ever since it was adopted for commercial purposes. In spite of its impressive safety record over the past 60 years, you can find some level of opposition to it everywhere it is used. It doesn't help that the development and use of nuclear energy have become, in many countries, political issues rather than scientific or technological ones. Decisions, therefore, are made for or against nuclear energy based on political sentiments to influence election results, for example, which is not beneficial to mankind's long-term well-being.

Nuclear energy has its pros and cons. It is one of the cleanest and most economic sources of energy. Many countries with limited natural sources have no feasible substitutes. And yet, nuclear radiation, once leaked in an accident, and nuclear waste, if not properly treated, could cause major environmental destruction and serious health problems to all those in the vicinity. While the likelihood of a large-scale nuclear accident is minimal, we should be aware of the risk. However, the remote possibility of a disaster should not cause us to give it up entirely. This would be tantamount to failing to see the forest for the trees.

Human Pollution

Life has many dangers. These include natural disasters, traffic and occupational accidents, environmental pollution, unsafe food, illness, the rise and fall of the stock market, and the vicissitudes of the economy. According to reliability analysis, nuclear power plants are very safe and nuclear energy is cleaner than most other energy sources. However, we can never be 100% sure that it is safe, and in fact, nothing is 100% safe.

In many cases, human activity is the main problem. In fact, coal has caused more human and environmental damage than nuclear energy. Coal burning is the prime cause of greenhouse gas emissions, which are directly or indirectly responsible for tens of thousands of deaths every year. It is the main cause of chronic illness and many other ailments. And yet, we never see protestors take to the streets to demonstrate against coal energy.

Background of Writing This Book

This book was written in the wake of the Fukushima nuclear accident of March 11, 2011. It aims to redirect the discussion on nuclear energy toward being a scientific and technological issue through some critical findings on a spectrum of seven categories of energy source. It also contains my reflections on what I saw during my trip to Japan as the first overseas nuclear safety scientist a month after the accident as well as on the reactions of people and governments regarding nuclear energy. In essence, the book looks at energy, environmental protection and safety, three vital issues of 21st century life, from a reliability point of view.

The original Chinese edition was published in Hong Kong in March 2012, on the one-year anniversary of the accident. The updated edition, greatly enriched, was published in both Taiwan and Hong Kong in May and July 2013, respectively. At the same time, the publication of the simplified Chinese, French and Japanese editions is under way. Other than those with translation citation sources, Longgen Chen performs the English translations of the Chinese poems used in the book.

Substantial supplementary materials are now included in the English edition covering the 100-year bloody history of coal-mining, the yet-to-be-solved air pollution problems, excessive CO_2 emissions, the extraction of shale oil, the prediction of the future of the global development of nuclear power by the World Nuclear Association (WNA), and the latest assessment of the nuclear accident in Fukushima.

Fukushima's Message to the World

On July 2, 2013, more than two years after the Fukushima nuclear accident, I was again invited to Japan to conduct a post-accident assessment on the safety of the Fukushima Daiichi Nuclear Power Plant. On the eve of the publication of this English edition of the book, I visited the Chernobyl site in Ukraine on October 3, 2013, some 10,000 days after the Chernobyl incident, to examine the area's recovery after the catastrophic nuclear accident that occurred on April 26, 1986 at the Chernobyl Nuclear Power Plant.

Although the World Health Organization (WHO) held that inhabitants living in the Fukushima area ran an increased risk of developing cancer, the United Nations Scientific Committee on the Effects of Atomic Radiation (UNSCEAR) put forward a different view.

After spending two years analyzing the situation, several dozen UNSCEAR scientists concluded in an interim report presented on May 31, 2013 that the amount of radiation that Fukushima inhabitants were exposed to was relatively low and that it was unlikely to threaten human health.

They even predicted that the nuclear plant workers who had been exposed to high levels of radiation were unlikely to face life-threatening risks or suffer from acute radiation-related diseases because they had taken proper preventive measures. Should these predictions prove accurate, there will have been no obvious radiation-related casualties among inhabitants living in the vicinity of Fukushima.

The Executive Committee of the European Union (EU) issued a mandate on June 13, 2013 asking for a compulsory examination of nuclear safety, demanding that the 132 nuclear reactors within the EU be subject to a legally binding review every six years and a safety check at least every 10 years. The Stress Tests report, *EU Nuclear Stress Tests: Legally Binding Reviews Every Six Years*, confirms the key point made in this book: the importance of transparency in nuclear power plant operations.

Although there is a disparity of views regarding the consequences of the Fukushima nuclear accident, there is no denying that so far two years after the accident nobody has died as a result of the radiation leakage in Japan. To some extent, this is a hard fact for many people to take in; it contradicts quite sharply with most media reports and online gossip.

Energy and environment protection are critical contemporary issues while the subject of reliability is a highly technical topic. We still have a lot to learn and much to improve. The Fukushima lesson drives home the message that we can no longer afford to be complacent or simply resign ourselves to fate in addressing the pressing issues of energy and environmental protection.

Transparency is the best guarantee of reliability. Insofar as safety and well-being are concerned, never echo the views of others without thinking. Most important of all, we should let the facts speak for themselves.

Way Kuo
City University of Hong Kong

January 1, 2014

Introduction

On March 11, 2011, Japan and the rest of the world were jolted by the Fukushima nuclear accident, triggered by a 9.0-magnitude earthquake and a serious tsunami and aggravated by human error.

After the incident, I was invited to comment on many official and unofficial occasions about the nuclear safety issues that caught the world's attention in the wake of the Fukushima nuclear accident: the hydrogen explosion in the nuclear reactor buildings, the discharge of radiation-contaminated water, nuclear radiation leaks, the impact these crises had on neighboring countries and their people and future energy policy including the use of nuclear power.

My comments and analyses drew wide attention from the mass media both at home and abroad. My viewpoints were widely reported in Hong Kong, Taiwan, the Chinese mainland as well as in the international press, even making it to the front page of Yahoo! News.

Visiting Sendai

On April 21, 2011, more than a month after the incident, I was invited as the first foreign scientist by the president of Tohoku University in Japan to give a lecture at Sendai about nuclear safety and reliability. I also made a trip to Tokyo and met with experts from Japan's nuclear supervising agencies.

At the opening ceremony in Beijing of the Seventh International Conference on Mathematical Methods in Reliability on June 20, 2011, I took the Fukushima nuclear accident as an example in introducing the concept of reliability design. This is a high-level international theoretical conference held bi-annually and the Beijing Conference was the first time that such a conference was held in Asia. The fact that I was invited as the keynote speaker at such a conference to talk about the Fukushima accident was therefore significant.

On September 9, I was invited to attend the 2011 Annual Convention of European Safety and Reliability Association (ESREL) at Troyes, France and deliver at its opening ceremony a keynote speech on the design of nuclear safety. Incidentally, a furnace explosion on September 12 near a radioactive waste storage site at Marcoule in southern France killed one person and injured four.

I visited Marcoule and found that the explosion was an industrial accident rather than a nuclear safety incident. Nevertheless, the media had seen it fit to use the word "nuclear" in their headlines, thus causing widespread anxiety.

Publication of Chinese Edition

In order to raise public awareness about energy sources and encourage a rational and fact-based debate, I wrote an article "The View of Nuclear Power from the Angle of Reliability" in April 2011, analyzing the development of nuclear power and its safety. Later, I published a series of articles in newspapers and magazines in the US, Hong Kong and Taiwan, which were widely reprinted by the international online mass media.

Meanwhile, I was invited to deliver lectures and give talks during 2011 and early 2012 at about 30 schools and institutions, where I shared my professional opinions on matters of energy sources and environmental protection. I also talked about related subjects at various universities in the mainland, Taiwan and Hong Kong. My lecture at the Chinese University of Hong Kong attracted an audience of 1,700.

In March 2012, to mark the one-year anniversary of the Fukushima nuclear accident that had drawn worldwide attention, I published in Hong Kong a book in traditional Chinese characters entitled *A Spectrum of Energies: Reflections on Energy and Environmental Protection in the Wake of Fukushima Nuclear Accident*.

The book was meant to promote public awareness of the interrelationship between environmental protection and energy sources. The fact that since April 2012, I received about 40 additional invitations proved that people were paying enormous attention to this important issue of the century. I have treasured and continue to treasure these precious opportunities to disseminate professional knowledge with the hope of promoting people's livelihood and well-being. However, when the memory of the accident fades, it seems people forget their concerns about the safety and reliability of nuclear energy as easily as the documents they have deleted from their computer.

Sustainability from the "Life Cycle" Point of View

Energy and environmental protection are modern-day problems that need to be addressed through interdisciplinary cooperation involving such disciplines as science and technology, economics, statistics, politics and psychology. People in general would like to have a simple answer as to whether something is environmentally friendly or safe. In fact, the assessment of the environmental friendliness and safety of a particular energy has to be undertaken by looking at the impact of the "life cycle" of an energy source.

Here's a simple example. It is the custom of American restaurants to serve guests routinely with iced water. It takes four or five glasses of water to wash a glass that has been used. Now, is it more energy-saving and environmentally friendly to use a glass or a disposable cup?

The answer depends on where you are. If you live in a water-deficient environment, it is more environmentally friendly to use a disposable cup, whereas it is more environmentally friendly to use a glass in a water-abundant environment. As a matter of fact, the majority of diners don't necessarily need iced water. As far as environmental protection is concerned, iced water should be served only to those who ask for it.

Let's look at some other examples. The use of corn to produce clean biofuel, for example, may aggravate the problem of food shortage. Solar energy is inexhaustible, but the production of solar panels consumes energy and causes pollution. With "unlimited state subsidy," moreover, it may result in the monopoly of the market, leading to international disputes. Wind energy is unstable and creates noise. Besides being offensive to the eye, wind farms pose a threat to birds and cause environmental pollution.

Hydroelectric power is clean and less polluting to the environment, but the fickle weather cannot ensure a steady supply of water. Besides, when building a dam on a trans-border international river, such as the Yalu Tsangpo River in Tibet, one has to take into consideration the interests of neighboring areas on the lower reaches of the river.

Fossil fuel generates 67% of the power consumed in the world, but coal-mining kills several thousand miners each year in China alone, to say nothing of the discharge of CO_2 that causes global warming. Coal cinders are fairly radioactive, a problem not to be neglected, but nobody seems to take heed.

Nuclear power is clean and cheap, but worries about its safety cause popular anxiety and endless arguments. Yet little progress has been made in the research and development of new energy sources.

Reliability and Safety

When one takes a panoramic view of human history, one will find that many issues and events go from being considered paramount to being disregarded.

In the field of energy sources, people may not overlook the potential danger of nuclear power, but they tend to overlook the important role reliability plays in the safety of nuclear power. They may not overlook the importance of energy sources, but they tend to overlook the necessity of nuclear power in human affairs. We tend to waver between sense and sensibility.

Only a few can accept the fact that there are at present no other appropriate and reliable energy sources to replace nuclear power. Some even think successful energy conservation can make nuclear power obsolete. They fail to recognize that energy conservation is always the prerequisite for the continuation of human life no matter what alternative energy we may adopt.

Choosing the right kind of energy is still an unavoidable step even when we succeed in conserving energy. Energy conservation is a necessary precondition to our existence, but is not sufficient by itself to obviate the need to search for safe and reliable energy sources.

As far as safety and reliability are concerned, people are accustomed to being reactive rather than proactive. Every time the launch of a space shuttle fails, I am invited to give my advice on whether the idea of reliability could help find a solution. As with the nuclear power plant accident, such requests come too late. We may complain about the way the spring blossoms and the autumn moon comes year after year, but when will we come up with proper preventative measures for tragedies such as the Fukushima accident?

Reliability is the key factor in maintaining balance between energy consumption and environmental protection. I'm certainly opposed to the irrational populist views that have cropped up since Fukushima; neither am I willing to let the present problem vanish like a cloud from the public's attention. If apathy were to cause another disaster, we would be left ruing our mistakes.

In the past year people have been plagued by a difficult decision about the development of nuclear power. Many countries hesitated over what move to take, as they reviewed their policy on energy.

Japan and US

Take Japan for example.

Immediately after the Fukushima accident on March 11, 2011, the whole of society, still in shock, was uncertain as to the future of nuclear power. Months later, a degree of composure returned, and as the scandal gradually came to light that both the government and the power company had neglected their duties, a voice continued to rise in opposition to nuclear power to such a point that nuclear power plants that were shut down for routine examination were not given permission by local governments to reopen.

It would appear that Japan was on its way to becoming a nuclear-free country.

During the summer of 2012 the Japanese government was surprised by the ensuing power shortages. In the panic, serious proposals emerged urging the resumption of nuclear power. Yoshito Sengoku, a veteran leader of the Liberal Democratic Party of Japan, even warned: "The shutdown of all the nuclear power plants is nothing short of a 'collective suicide' for Japan". He sounded more frightened by such a prospect than the 9.0-magnitude earthquake itself. One can easily imagine the vehement opposition that his remarks aroused.

When summer came, the number of old people who died from the heat due to a lack of air conditioning sharply increased in Tokyo, causing anxiety and concerns and highlighting the fact that there is a price to pay for having a limited supply of electricity.

The Liberal Democratic Party of Japan regained power after winning a historic victory in the House of Representatives election on December 16, 2012. The Tokyo Electric Power Company (TEPCO) led the stock market in price hikes, and it was to be expected that Japan's new government would soon resume the use of nuclear power.

On March 4, 2013, a few days before the second anniversary of the Fukushima nuclear accident, US President Obama nominated E. J. Moniz of MIT as his Energy Secretary. Moniz strongly advocates the policy of reaching the goal of environmental protection and carbon emission reduction by developing nuclear energy.

All these have once again proved the main assertion of my book: a balanced relationship must be kept between the three aspects of energy supply, economic well-being, and reliability and sustainability.

Technology, Energy Conservation and Environmental Protection

It is worth mentioning that I was invited to give key lectures at three different conferences in June 2012.

The first one came on June 11 at the annual conference of the Chinese Academy of Engineering in Beijing where I gave a keynote speech on the balance between energy and environmental protection to a gathering of about 1,000 academicians and professionals. On June 14, I delivered a keynote speech on the theoretical basis of energy safety and industrial safety design at the 2012 International Congress of Applied Probability in Jerusalem. On June 23, after the opening ceremony of the Second International Conference on the Interface between Statistics and Engineering (ICISE) at National Cheng Kung University of Tainan in Taiwan, I gave a keynote speech on the assessment and application of nuclear safety.

In addition, on July 21 and August 1, I delivered a general education lecture on the theory and practice of reliability at the 2012 Hong Kong Book Fair and at the Tenth Asian Network for Quality Congress held at Hong Kong University of Science and Technology, respectively. On August 8, I was invited to visit No. 4 Nuclear Power Plant in Lungmen, Taiwan, which had not yet come into operation, and talked about the relationship between the practice of reliability and nuclear safety from the angle of internationalization.

The modern concepts of energy management and environmental protection are highly complicated and professionalized topics. Empty rhetoric about energy conservation and environmental protection is no help.

We need to formulate a comprehensive policy on energy conservation, environmental protection, innovation and safety of energy use. In addition, we should constantly develop new energy sources and design an optimal plan of power distribution by making use of smart grid and cloud computing technology. In many cities in Europe, marked progress has been made in the development of low-carbon ecological cities, especially in Scandinavia where outstanding improvements are particularly praiseworthy and can serve as a reference point for other places.

In order to fulfill our responsibilities of promoting the well-being of the world's population and making contributions to the world, we have to analyze the issues of energy and ecology in a multi-faceted way. For example, for the purpose of ensuring an ecological balance and a sustainable energy operation, we can follow the example of Europe where each country has its smart grid connected with that of the neighboring countries. Furthermore, many countries and regions are required not only to develop new energies, but also to actively promote the connection of its smart grid. In this way we can benefit from a spectrum of highly efficient, safe, low-carbon, economically optimal energies.

Rainbow Energies: a Spectrum of Energy Sources

Energy shortage is a pressing issue that faces all mankind in the 21st century. In my opinion, there is a whole spectrum of seven different kinds of energy (I call them "rainbow energies") such as hydropower, thermal (coal, oil & natural gas), nuclear, wind, solar, biofuel and others (geothermal, ocean energy and marsh gas).

All have strengths and weaknesses in terms of efficiency, safety, reliability, environmental protection, reserves and economic value. Some of the shortcomings are causes for legitimate concerns, but others are not causes to worry, and sometimes subject to political manipulation.

In evaluating different energy sources and making energy policies, we must adhere strictly to the principle of taking into consideration all factors such as environmental protection, "life-cycle" pollution, the reliability and sustainability of the energy sources, and the financial and non-financial risks that people have to face.

In a word, in order to address the problem of energy and environmental protection, we need to pool together the wisdom of different disciplines and carry out inter-domain cooperation. On the principle of sustainable development, we shall promote people's well-being, without forgetting to remain master and not slave to the material world.

Crisis Management

The Fukushima nuclear accident reminds us again that human history is often written and illuminated by one crisis after another. In the areas of energy conservation and energy innovation, there are still considerable improvements to be made.

A crisis is symmetrical by its nature: any decision-making is accompanied by a crisis. Even refraining from making a decision comes with its own kind of crisis. Therefore, the popular saying that "the more one does, the more mistakes one makes; the less one does, the fewer mistakes one makes; only when one does nothing will no mistakes be made" does not hold water. Otherwise, why should we talk about innovation?

Publication of "Critical Reflections on Nuclear and Renewal Energy"

The first Chinese edition of this book was published by Hong Kong Cosmos Books Ltd, which also published an updated Chinese edition in July 2013. In addition, a Taiwan edition was published in April 2013 by the Taiwan Commonwealth Publishing Group, while the publications of a simplified Chinese-character edition, a French edition, and a Japanese edition are in progress.

On the basis of the Chinese edition, the updated Chinese edition has been enriched with a wealth of materials, especially regarding policies that concern science and technology, energy sources, environmental protection and occupational safety.

Years of observation have convinced me of the need to deepen the general populace's understanding of the concept of reliability.

Addendum

The transcripts of two recordings make up the appendices: Appendix I is revised from a recorded speech of mine about reliability and its historical development. Appendix II is taken from a conversation about the Fukushima nuclear accident and the ensuing crisis with Hong Kong's noted journalist Chip Tsao at the 2012 Hong Kong Book Fair.

Excerpts of two English reports published at the end of 2011 are included in the book. One is about Bill Gates' attempt to find a partner in China to build a new-generation "travelling-wave nuclear plant." The other is a special report issued by the Institute of Electrical and Electronics Engineers (IEEE) in the US chronicling in detail the events in the 24 hours after the Fukushima nuclear accident and its bitter lessons.

In addition, I have included in the Postscript excerpts of an important report written in early July 2012 by a fact-finding team of 10 experts appointed by the Japanese government. Their conclusion coincides with the viewpoints I expressed in the Chinese edition published in March 2012.

Some materials presented in this book were published previously in *Ming Pao Monthly*, *Hong Kong Economic Journal*, *The South China Morning Post*, *AM 730* (for the excerpted dialogue) in Hong Kong, *IEEE Transactions on Reliability*, *IEEE Reliability Society Newsletter*, as well as *China Times* and the National Science Council in Taiwan. I would also like to express my thanks to Longgen Chen of City University of Hong Kong, who has done a lot of work for the publication of this book.

All royalties from the Chinese edition are donated to student scholarships.

Part One

The World after March 11, 2011 — Ripple Effect of the Fukushima Accident

1

Reliability and Nuclear Power

A 9.0-magnitude earthquake and the tsunami it triggered hit the north-east coast of Japan on March 11, 2011, causing widespread death. The destruction associated with this event included successive hydrogen explosions in three reactors and one spent fuel tank at the Fukushima Daiichi Nuclear Power Plant.

The released radiation triggered a backlash against nuclear power in several countries. Some overseas rescue teams abandoned Japan for fear of radiation exposure. Some people in neighboring countries scrambled to buy iodine and iodized salt for fear of nuclear pollutants blowing across from Japan.

Although people in Hong Kong for the most part knew there was only a slim chance that radioactive substances could reach the territory, many rushed nevertheless to the supermarkets to purchase Japanese milk powder and soya sauce, fearing future supplies might be contaminated.

As a rule, any disaster is likely to cause panic, and this is particularly so with nuclear energy-related crises. When I first came to City University of Hong Kong in 2008, I delivered my first academic talk on nuclear safety. But only a few people paid heed.

With the nuclear crisis in Japan, many people are now gripped with fear, reminding me of the Three Mile Island accident in the US in 1979. At that time, Americans panicked, too, in no less degree than people are doing today. The accident this time was more grave than the Three Mile Island accident. On April 12, a month after the accident, the Japanese government raised the crisis level to seven. The fear is understandable since fear is generally of the unknown and most people don't know much about nuclear energy.

Nuclear Accidents in History

Nuclear energy is a cross-disciplinary science and technology.

The world's first nuclear power plant for civilian use was opened in Obninsk in the former Soviet Union on June 27, 1954. Its output capacity was 5 megawatts (mW) of electricity. The nuclear power generating unit that opened at Windscale, UK on August 27, 1956 was the world's first commercial nuclear power plant. Up until 2012, the operation of nuclear power had a history of about 60 years. As of March 11, 2011, there were 434 nuclear reactors in over 30 countries as indicated in Figure 1, providing about 13% of all the electric power consumed in the world for that year. Japan now possesses 50 nuclear reactors, ranking the nation among the highest in terms of the utilization of nuclear power. These reactors provided 30% of the total national power supply before the 2011 Fukushima accident. The top-ranking country is France, with over 75% of its power generated by nuclear plants. See Figure 1 for World Nuclear Reactors (2013) and Nuclear Electricity Generation (2011).

There have been several serious nuclear accidents in the past, the most critical being the Three Mile Island nuclear accident in the US in 1979, the Fukushima accident in Japan in 2011 and the Chernobyl accident in the former Soviet Union in 1986. The last is the only nuclear accident to date to inflict heavy human casualties. It should be noted, though, the design and construction of the Chernobyl plant was far from compliant with international standards and it should not be counted as a standard nuclear power plant.

In Japan, there had been three incidents in the nuclear power industry prior to Fukushima since the first plant opened in 1966: a leak of radioactive wastewater at Tsuruga nuclear power plant in 1981, an incident at Monju nuclear reactor in 1995, and the critical accident at JCO Fuel Fabrication Plant in Toki-mura in 1999. None inflicted serious casualties on civilians.

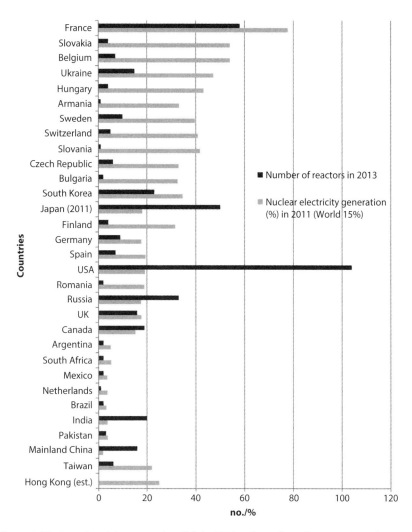

Figure 1 Nuclear electricity generation (%) in 2011 and number of reactors in each country in 2013.
Source: World Nuclear Association
http://world-nuclear.org/info/reaactors/html

Nuclear Power Plants are not Atomic Bombs

One of the chief reasons behind the panic comes from a fear of atomic bombs.

Before the end of World War II, two US atomic bombs were dropped on Japan. The ensuing devastation left an indelible shadow and bitter memories. However, as the fire in the stove at home is different from burning grass on waste land, and firecrackers are different from hand grenades, a nuclear power plant is not an atomic bomb. The uranium concentration in a nuclear power plant is not high; it is confined within the solid, unventilated furnace where a controlled nuclear reaction takes place. What's more, the regular cement wall outside the nuclear reactor is two- to three-meters thick. In contrast, the uranium concentration in an atomic bomb is high and hard to control.

What worries people most in Japan is radioactive substances contaminating water sources. So far, this has not happened, but in case it does, there is no reason for panic as people living nearby have been evacuated. The negative effects would be far less devastating than those caused by an atomic bomb.

Is Nuclear Power Reliable?

The use of nuclear energy to generate electricity has been controversial ever since the first power stations opened, but the Fukushima nuclear accident has brought these concerns sharply into the public eye. People are anxious to know more about the chances of similar accidents occurring in the future and the potential impact. So is nuclear power reliable enough?

It should be noted that the nuclear industry adheres to strict safety standards. In 1975, *WASH1400*, also known as the Rasmussen Report, was issued in the US. When I was a graduate student, I read with great interest the accounts of all the possible scenarios for accidents, the probabilities of such accidents and a detailed analysis of possible reasons behind the accidents. In 2007, the Nuclear Regulatory Commission (NRC) published its State-of-the-Art Reactor Consequence Analysis (*SOARCA*), which uses computer simulations to illustrate potential accidents in nuclear power plants, and has become the universally-accepted guideline for nuclear power safety.

Reliability is a mode of assessment. In manufacturing industries, reliability maintains a close relationship with the product warranty. It includes elements such as product-cycle analysis and risk evaluation. As far as nuclear power is concerned, reliability is achieved by assessing every link in the whole process for possible failure, similar to how and what *SOARCA* records.

Nuclear power is a mature industry now. Both Generation II and III nuclear power reactors are built according to sound scientific theories and safe technology. The probability of a nuclear accident is minimal. Physicist B. L. Cohen has concluded in a study on life expectancy (Appendix I) and radiation risk from nuclear power plants that, even if a person lived his or her entire life in the vicinity of a nuclear power plant, the impact on life expectancy would be far less on average than that due to non-nuclear-related accidents. (See Table 1 below.)

"Second-hand Contamination" — An Unfounded Notion

Only people living close to the source of a radiation leak from a nuclear plant are in danger. Nuclear radiation harms the human body only when it accumulates to a certain level. In the event of an accident with off-site risk, people should be evacuated outside a 30 km radius.

Medical treatment for radiation sickness is required for anyone who stays within that 30 km radius for several hours. The consequences of exposure to high levels of radiation include infertility, thyroid carcinoma and other cancers, depending on the levels of exposure.

The fact is that small amounts of radioactive substances attached to shoes and clothing can be washed off easily, and will not pose any harm to human health.

Table 1 External factors affecting life expectancy.

External factors	Days of lost life expectancy
Male	2,800
Smoking a pack of cigarettes daily	1,600
Poverty	700
Accidents	435
Occupational accident	74
Drinking a can of soft drink daily	2
Air travel	1
Living near nuclear power plant	0.4
All US energy from nuclear power (NRC)	0.03

Source: B. L. Cohen, Before It's Too Late.

The Three Mile Island nuclear accident was rated a five on a seven-point scale, according to the International Atomic Energy Agency (IAEA). Tracking reports by the NRC and several other independent research reports suggest that the highest background radiation dose was between 100–125 mrems, and the accumulated radiation that the 2 million residents in the vicinity were exposed to was similar to the average natural radiation that we are exposed to every year. (We should note that radiation from an X-ray is 6 mrems). These reports suggest that Japan's neighbors need not worry about radiation leaking from the nuclear power plant.

Nuclear Power Plants: Economical and Safe

Rapid economic development has aggravated the global energy crisis. Every country is looking for alternative energy sources, of which nuclear energy is a top priority for many. Under current conditions, nuclear power still boasts unrivaled merits and is the most economical even considering the dismantling cost and cleanest power source.

Nuclear power and coal-derived fuel rank among the most economical forms of energy. The average is about $0.022 per kilowatt-hour (kWh), less than half the cost of petrol. However, according to statistics issued by the World Nuclear Association (WNA), China spends 6% of its GDP every year dealing with pollution caused by burning coal and oil.

In addition, a nuclear power plant is one of the safest workplaces you can find. The probability of a workplace accident is higher in the finance, insurance or real estate development sectors. For every 200,000 working hours, the ratio of an accident in a nuclear power plant is less than one versus the average of 45 in the general manufacturing industry.

According to WNA statistics, prior to the Fukushima nuclear accident, only two severe nuclear accidents, the Chernobyl and Three Mile Island accidents, occurred throughout the combined total of 14,000 operating years of commercial nuclear reactors in 32 countries.

In addition, viewed in terms of energy value, nuclear power generates less harmful substances, such as CO_2, than other energy sources do. In fact, the environmental value of nuclear power is far higher than that of other energy sources, as illustrated in Table 2.

Presently, the generation of nuclear power has incomparable advantages and has proved to be a highly economical and clean source of electricity. In terms of reducing carbon emissions and protecting the environment, no other energy can surpass nuclear power.

Table 2 Highest environmental value of different energy sources.

	NOx	SOx	Hg	CO$_2$	Total value	Present value
	($/kW/year)					
Nuclear energy	11.6	5.3	12.0	61.2	90.1	750.1
Wind energy	2.2	0.7	1.6	17.1	21.6	180.1
Solar energy	1.5	0.3	0.8	13.7	16.3	135.3
Bioenergy	−4.3	3.7	8.5	51.5	59.4	495.1

Source: Electric Power Research Institute, 2003.

Monitoring Imported Japanese Food

Based on a close examination of a number of factors, concerns over food imports from Japan, while understandable, are not justified. There are safety checks on all imported food from Japan using special detection devices for radioactivity, just as there are X-ray checks at the airport on luggage.

If the fuel in the stricken reactors has not melted, however, these checks are unnecessary. On this point it is important to keep in mind that radioactive substances cannot produce any negative effects without a medium. While "extremely low" levels of radioactive iodine have been detected in the air over many neighboring countries and regions, they are unlikely to pose any threat to public health and the environment.

In any case, a lot of food is subjected to small doses of radiation as part of the overall manufacturing process. Food irradiation (exposing food to ionizing radiation) is widely used for sterilizing, preventing decay and killing insects. The milk we drink and many of the green vegetables we eat have been subjected to similar treatment.

Human Factors

Even if nuclear power plants are built on a sound scientific and engineering basis, the biggest uncertainty is the human factor. Both the Chernobyl and Three Mile Island nuclear accidents happened at midnight and were caused by human error. Paying attention to the interaction between humans and systems is crucial to safety in all kinds of systems. In a very big system, we must locate bottlenecks and implement radical reforms where necessary. Professional training is an ongoing process, ensuring staff members are well-trained is a better guarantee of safety than reliance on high-tech equipment.

It is of paramount importance to remain calm in the face of an imminent crisis. Getting to know more about nuclear energy will enable us to make sound judgments in such circumstances. However, professional knowledge about nuclear power is usually something beyond the reach of the average person, and this is the cause of the panic. Therefore, while the nuclear industry and nuclear power plants need to enhance safety measures, they should also promote popular science education on nuclear energy.

Paying Heed to Maintenance

We should also increase the transparency of the operation and management of nuclear facilities to dispel the psychological pressure on the general population. As we know, panic in certain sectors of the local community after the nuclear accident in Japan had much to do with the lack of perceived transparency. As a result, people questioned the credibility of the released information, creating room for speculation and anxiety. People prefer to err on the side of caution.

The Fukushima nuclear plant has been running for about 40 years, its designed limit. It remains to be seen whether there was any negligence in the maintenance strategy prior to the 2011 disaster. We are now more keenly aware than ever before of the importance of vigilance and maintenance.

Using car safety as a metaphor, we expect to use spare tires at some point when we are driving. If dozens of nuclear reactors are running simultaneously, there should be several spare systems and power provision centers ready to serve as replacements at any time. Constant checks should be conducted on management, operation and contingency measures to ensure smooth running.

Nuclear Waste and Waste Management

Radioactive byproducts created by the generation of nuclear power, as well as solid and liquid substances that have been exposed to nuclear radiation, are a cause of social unrest. This nuclear waste can be classified according to radioactive intensity as low-level, intermediate-level or high-level.

Low-level radioactive materials include tools, waste water, garments, shoes and caps, plus any other components and parts that have been exposed to radiation in the nuclear power plant.

Intermediate-level radioactive materials are normally produced in the process of nuclear reaction. As intermediate-level nuclear waste is highly radioactive, its treatment requires the shield of a cement wall or protective clothing for workers. Radioactive iodine and cesium released into the air after the Fukushima nuclear accident were intermediate-level waste.

Spent fuel containing high-level radioactive isotopes after nuclear fission constitute high-level nuclear waste.

Physicists have continuously researched in dealing with the nuclear waste problem. One prospective approach is to develop a fusion generator that could make the fuel rods removed from the nuclear power plants safer to store. If successful, this could help solve the nuclear waste problem and at the same time give fusion research a new *raison d'être*.

A typical nuclear power plant may produce some 100 cubic meters of low- to intermediate-level nuclear waste each year. Such nuclear waste should be isolated from the outside world for 200 to 300 years. Low- to intermediate-level radioactive materials are wrapped up and buried under shallow ground (near surface disposal).

Because high-level radioactive waste contains a small quantity of ultra-high radiation materials with a very long half-life cycle, it should be vitrified and wrapped in multi-layers after cooling for a long time and then buried in solid rock deep underground (deep geological disposal) to isolate it from the outside world for thousands of years.

Generally speaking, a nuclear power plant produces two metric tons of high-level nuclear waste each year.

Low-level radioactive nuclear waste, particularly radiation-polluted waste liquid, comprises a large proportion of the total nuclear waste, which can be solidified first of all, to significantly reduce its volume before it is stored in dry casks.

Though of a small quantity, the spent fuel generates a high level of heat and is highly radioactive. It is therefore usually stored in ponds at reactor sites for several years until it cools down. It is then processed for long-term dry storage.

Ever since the 1970s, many countries have devoted a great deal of effort to properly treat nuclear waste, with an eye to exploring and developing safe and reliable techniques and manufacturing dry-storage equipment in the nuclear power plant. Storage techniques have matured and are now applied in many countries. In addition, techniques for the final treatment of high-level nuclear waste have improved a great deal.

Over 30 countries using nuclear power across the world have worked out different timetables for the final treatment of nuclear waste. Finland will be first, putting to use the equipment for the final treatment of nuclear spent fuel in 2020; France and the US are expected to have their arrangements ready in 2025 and 2045, respectively.

At present, in addition to burying high-level waste deep in underground rock, spent fuel may be transported to Europe and the US for further treatment, so that plutonium may be extracted as a new type of fuel for generating electricity.

Since nuclear waste is highly radioactive, the latest research is focused on new re-treatment techniques that will burn up nuclear fuel completely. In this way, the utilization ratio of nuclear fuel is greatly raised on the one hand, and the production of nuclear waste is greatly reduced on the other. This means that nuclear waste can't be used to make nuclear weapons or dirty bombs for terrorist activities.

Impact of Nuclear Science and Technologies on Modern Society

Nuclear power constitutes 25% of Hong Kong's total power supply, and is expected to increase to 50% by 2020 to help curtail the release of CO_2. As it is impossible to depend on any alternative power supply within a short period, banning nuclear power would mean limiting the use of electricity and intermittent blackouts, gravely impacting life in Hong Kong. Needless to say, we should investigate developing technologies for renewable energies and advocate a new energy-efficient lifestyle.

But nuclear energy still ranks among the major energy sources after taking into account all the factors, including the negative impacts of alternative energy technologies and cost, the pollution of non-nuclear power to the environment, the accumulated experience derived from accidents, breakthroughs in science and technology, new advances in industrial techniques, etc.

I believe that, learning the lessons from Fukushima, the people of large cities such as Hong Kong have come to see that they are not immune to changes in neighboring countries and regions. In a globalized environment, what happens in the region will affect the local economies as well as political and social stability.

In modern society, we need to know more about science and technology and promote scientific and technological advancement to solve the problems confronting humanity rather than fear science and technology due to a lack of understanding. Otherwise, we will not be able to enjoy the high quality of life that comes with science and technology.

Social and National Security

As a popular saying goes, "A vigilant person will net the rarest cicada; a cautious person can sail safely all the way". However, vigilance and caution are just two of the preconditions for safety. In present-day society, maximizing people's well-being and safety requires another, and more important, precondition: innovation.

The Chinese word *anquan* means both "safety" and "security", and the distinction between the two concepts is thus lost when they are rendered in Chinese. Chinese people almost always think of *anquan* in the sense of "safety," and that is how I use the word in most cases in the following chapters. When the same word is applied in the discussion of the energy spectrum, however, it carries the additional meaning of social and national security, a point that has unfortunately drawn little public attention.

The "Spring Scream Music Season" (generally known as "Spring Scream") was held on April 4, 2013 near the No. 3 Nuclear Power Plant in Kenting, in southern Taiwan, arousing concerns about nuclear security. The Taiwan Power Company assured that the routine operation of the nuclear power plant would not be affected by Spring Scream and that manpower deployment would not be a problem. The organizer of the Music Season also expressed confidence that Taiwan was capable of handling the security of the power plant.

Besides, the Music Season was going to be held over a weekend.

There was no need to be overly concerned.

However, suppose some terrorists were to invade and occupy the nuclear power plant, wouldn't you agree this would be a cause of concern for social security?

Or suppose some minor mishaps were to occur to the No. 3 Nuclear Power Plant on April 4 and rumors were to spread among the thousands of concert attendees, and suppose the crowds did not disperse as planned and the rescue and support teams failed to arrive in time, what do you imagine the outcome would be?

Similar scenarios are not difficult to imagine, and the chance of their happening is often higher than is generally recognized. How else is one to explain the strict security checks at places such as the airport? Moreover, according to a report in the *Sankei Shimbun* on May 29, 2013, a former officer of the People's Army of North Korea revealed that it made plans to launch suicidal terrorist attacks against nuclear power plants in Japan and South Korea.

On September 11, 2001, terrorists used civilian airplanes to attack the New York World Trade Center in the US, killing at least 3,000 people. In the five months following the attacks, wildly toxic dust from the pulverized buildings continued to fill the air of the World Trade Center site. As of September 11, 2013, at least 1140 rescue workers and local residents have been diagnosed with cancers, particularly prostate cancer, thyroid cancer, and multiple myeloma, due to their exposure to the many toxic dust and fumes including Dioxins and Polycyclic Aromatic Hydrocarbons (PAHs).

Since then, various governments have heightened their vigilance and introduced emergency measures to provide greater security to their people and protect the infrastructure vital to the nation and society.

A major factor behind the wars in the Middle East in 1973, 1979 and 1990 was oil. That's why they were called the wars caused by the energy crisis. As electric power and fuel energies have a direct bearing on people's livelihood, the issue of energy is far from simply a resource problem, and will undoubtedly become a major challenge to national and international security and social stability in the 21st century.

From a similar perspective, wouldn't the shortage or disruption of the electricity supply that resulted from a failure to diversify energy sources or to connect smart grids threaten national security? Conversely, if Taiwan were to suspend its nuclear power without any safe environmentally-friendly alternatives to take its place, making it necessary to connect with the Fujian nuclear power plants across the Taiwan Strait, how would people in Taiwan feel, then?

2

Some Flowers Fall, and Again They Bloom

Japan is famous for its cherry blossom, and every spring there is a buzz of excitement in anticipation of the picturesque scenery throughout the country.

But 2011 was very different. In the days between the departure of the cold winter and the advent of a chilly spring, Japan was hit by an unprecedented 9.0-magnitude earthquake and an ensuing tsunami.

The earthquake and tsunami inflicted heavy casualties in the Tohoku region of Honshu, the main island of Japan. An enormous number of people died and a great deal of property was destroyed. The televised images of giant waves devouring lives, destroying houses, carrying away vehicles and submerging an airport were broadcast around the world.

But perhaps what held people's attention the most was the stricken Fukushima Daiichi Nuclear Power Plant when thousands had to abandon their homes while the fallout from the disaster was being assessed.

Now that so many countries are keen to develop nuclear power as an alternative energy source, safety has become a hot topic. Successive reports about the crisis at the nuclear power plant held the public's attention while the news of non-nuclear related casualties – with nearly 25,000 dead or missing - seemed to sink into oblivion. The destruction of the oil refineries was almost entirely ignored.

Visit to Sendai During the Cherry Blossom Season

On April 20, I delivered a talk about the future prospects for nuclear power among experts at Tohoku University and scholars from Tokyo during a three-day visit to Tohoku University, which is situated in Sendai, the closest city to the epicenter. The name of the city comes from a poem composed by Han Hong of the Tang Dynasty, entitled "Jointly composed at the Lodge of Roaming Immortals". *Sendai* in Chinese means "Immortals' Terrace".

> *Below the immortals' terrace one sees the towers of the Five Cities,*
> *things in the scene are dreary, the nightlong rain withdraws.*
> *The colors of hills stretch far to reach evening in the trees of Qin,*
> *fulling blocks' nearby sounds tell of autumn in the Han palace.*
> *Shadows of a sparse pine fall on the calm of the empty altar,*
> *scent of tiny plants opens into the seclusion of a small grotto.*
> *What use to seek out elsewhere, go off "beyond the norms"?*
> *in the world of mortals also there is a Cinnabar Mound?*

(Translation quoted from: *The late Tang: Chinese Poetry of the mid-ninth Century (827–860)*, by Stephen Owen)

Right after the disasters, I visited Sendai, which is the capital of Miyagi prefecture and the largest city in the Tohoku region. More than 20 years ago, I visited Tokyo on a number of occasions and had close contact with the then-Ministry of International Trade and Industry (MITI) and was at one time on their recruitment list. Now, MITI has been changed to the Ministry of Economy, Trade, and Industry (METI). My return visit was filled with a different feeling.

I had to take the Shinkansen (the high speed bullet train) from Tokyo to Fukushima and then transfer to Sendai as Sendai Airport was still closed to international flights due to the flooding. But my Japanese host told me when I arrived at Haneda Airport in Tokyo on April 19 that even though international flights from Sendai Airport were still suspended, domestic ones had resumed. Not only that, the Shinkansen from Tokyo to Fukushima was up and running just one month after the earthquake and tsunami. The whole line would return to normal the following day, i.e. April 20, he said. Such recovery speed was really admirable.

As winter gives way to spring, Sendai usually sees a seasonal revival as tourists flock to sample the inviting sights, sounds, seafood and vegetables, and experience the famed Japanese sense of serenity, safety and order. Throngs of visitors usually flock to feast their eyes on the beautiful cherry blossom.

However, many Hong Kong and Taiwan tourists, for instance, stayed away this year, hurting local businesses and tourism. Tohoku University put me up during my visit at a local *minshuku* because a lot of hotels in the vicinity had not yet fixed their damaged interiors. The *minshuku* host was very hospitable, which was so different from my usual impression of the conventional image of the Japanese, as if he was trying to restore life back to normal and get rid of the ill effects of the earthquake.

Whether in Sendai or Tokyo, where I later visited, there were several aftershocks every day, but life went on as usual, even though there were restrictions on the use of electricity in Tokyo's subway system and elsewhere. No one seemed unusually ruffled in the capital.

Of course, the cherry trees bloomed from the south to the north of the country when spring arrived. The only difference was the absence of laughter under the trees from admiring visitors. The sights seemed to echo the melancholic beauty that Wang Ling, a poet of the Song Dynasty, described in his poem, "Bidding Farewell to Spring":

> Spring's last month sees flowers fading, they fall, again they bloom;
> And little eaves find Swallows arriving daily back to their room.
> Into the midnight the Cuckoo wails, weeping till she bleeds,
> Convinced her crying never fails to regain the Spring's breeze.

Tohoku University

We associate Japan with quality, and the prestigious Tohoku University is no exception. It ranks among or close to the top 100 universities in the world in the *Times Higher Education* World University Rankings, Quacquarelli Symonds, and Shanghai Jiao Tong University's Academic Ranking of World Universities. No other university in China, Hong Kong, or Taiwan enjoys the recognition Tohoku enjoys. Tohoku also ranks 1st in Asia in the Broad Subject Field of Engineering/Technology and Computer Sciences by Shanghai Jiao Tong University's 2013 Academic Ranking of World Universities.

But it is not a boastful institution. Like the University of Tokyo, Kyoto University and other famous universities in Japan, Tohoku does not have an imposing gate or ostentatious sign. However, its academic standing is anything but mediocre. It boasts not only Nobel laureates among its faculty and alumni but also nominees among its existing colleagues and peers, including its President, Professor Akihisa Inoue.

One of those alumni Nobel laureates is Mr Koichi Tanaka, co-winner of the Nobel Prize in Chemistry in 2002. A 1983 graduate, Mr Tanaka worked on the development of mass spectrometers at Shimadzu Corporation even though he doesn't have a doctoral degree. Coincidentally, neither does my colleague at Texas A&M University, Mr Jack Kilby, the winner of the 2000 Nobel Prize in Physics.

In my view, a degree is not the same as scholarship, a point I repeat when I talk to young people. There is no exception in scientific and technological circles, let alone in business and entertainment. To further extend my argument, having scholarship does not necessarily mean one is able to administer or rule the world. The only way lies in carrying the truth and applying learning to practice.

We often, quite rightly, associate age with wisdom, but the age of a university should not be used to judge its performance. Universities in Japan are generally much younger than those in Europe where 400- or 500-year-old universities are not uncommon, or those in North America which can be 200 or 300 years old.

The University of Tokyo, the oldest in Japan, is relatively young—only 134 years old—while Tohoku University was founded in 1907, making it younger than Peking University in China, which was established in 1898. Interestingly, in 1904, when Lu Xun, the well-known Chinese scholar, pursued his studies at Sendai Medical Academy, the predecessor of the Medical College of Tohoku University, the buildings for Tohoku University had not yet been completed. Despite their relative youth, compared to the historic European universities, Japanese universities have contributed greatly to Japan's development since the Meiji Restoration in the mid-19th century.

In addition to efficiency, modesty and quality, Japan's professors are known for their scrupulous scholarship. When I turned to a veteran professor for his views on the earthquake of March 11, he replied that the day had "finally come". What he meant was the Japanese, known for their thoroughness and meticulous attention to detail, had held repeated rehearsals in preparation for the earthquake.

When I was leaving Sendai, spring rain was lashing down on the cherry blossoms growing on the hills. The sight reminded me of lines in "Fallen Blossoms" by Zhu Shuzhen, a poet of the Song Dynasty. She wrote:

> *The blossoms on the twin trees in full bloom;*
> *The jealous winds and rains shake them and toss.*

Would that the King of Spring be e'er in reign!
Don't let the blossoms drop and dot green moss.

(Translation quoted from the following web site:
http://www.yidian.org/articlelist.php?tid=16115)

I feel sure that with the advent of the next spring the goddess of flowers will work her wonders again, transforming Sendai into its former glory. I sincerely hope that the teaching and research at Tohoku University will flourish once again just like the exuberant cherry blossoms, and that the green moss will not blight its beauty.

Starting from this chapter, I am going to present a series of reflections concerning my visit to Japan in April 2011. I will analyze from a scientific perspective how we can make use of nuclear energy to promote economic development and reinforce safety. This series of chapters will conclude by the time the spring flowers are gone and the autumn moon rises. By that time, I hope Japan will have brought the nuclear accident under control. It is my sincere hope that we will benefit from a critical analysis of the causes and effects of the disaster, learning lessons that will help to promote social advancement.

3

Different Responses Across the Waters

The Tokyo Electric Power Company (TEPCO) aggravated an already tense situation in Japan following the shutdown of the Fukushima Daiichi Nuclear Power Plant earlier in 2011 by suppressing the full story behind the radiation leak out. The situation was not helped by a rigid system for responding to the crisis and insufficient contingency plans.

Despite these trying circumstances, the Japanese remained calm. There were no reports of panic buying in supermarkets or social disorder on the streets. Against a backdrop of tens of thousands dead, missing or injured, and unrelenting attention from the global media, the rescue teams and the public worked together to get society back on its feet.

The rational way the Japanese confronted these disasters and other problems should be applauded. Rationality is a worthy attribute, as we can see from the following survey. A month after the nuclear disaster, the *Asahi Shimbun* polled the public on nuclear power. These are the results, Table 3, compared with responses taken in 2007.

Table 3 Changes in Japanese attitude towards nuclear power.

	2007	April 18, 2011
More nuclear power	13%	5%
Acceptance of nuclear power	53%	51%
Less nuclear power	21%	30%
No nuclear power	7%	11%

Note: A similar poll conducted in April 2011 in the US shows 58% support nuclear power.

The figures make for interesting reading. They show that, even though Japan is the only country in the world to have had nuclear bombs dropped on its soil, a certain proportion of its population is still ready to accept nuclear power. We see that the number of people in favor of reducing nuclear power has increased by 9%, while the number that accepts nuclear power remained at over 50%.

On January 6, 2013, Reuters reported a survey of 135 mayors in cities in Japan located around the 50 nuclear reactors which had ceased operation. In the survey, which was conducted by Japan's *Yomiuri Shimbun*, 54% of respondents supported the restart of their respective nuclear power plants on condition that assurances about nuclear safety were offered. Around 18% of respondents were against a restart. These results suggest to some extent that people have to face reality when economic well-being is at stake.

We should take note that the Japanese tend to reach decisions through balanced thinking. They, more than any other nation, are aware of the dangers of nuclear power, being the only country that has been hit by atomic bombs. But they have weighed up the pros and cons and come out generally in favor of nuclear power. By no means is the public in total agreement. But the people have sufficient knowledge and full recognition of the reality to make an informed choice.

Their rational thinking is also evident in a poll indicating that 65% of the population agrees with the implementation of a consumption tax, an issue that divides societies elsewhere around the world.

However, according to the report on June 6, 2012 in *Japan Today*, a poll conducted in 2012 by the Pew Center in the US shows that one year after the Fukushima nuclear accident 80% of Japanese people are dissatisfied with the effect of their government's handling of the nuclear crisis. As a result, 70% of the public think Japan should reduce its dependence on nuclear power. In 2011, only 44% held the same opinion.

In the meantime, a survey carried out by a Japanese TV station demonstrated that about 70% of Japanese people chose "don't know" as their answer to the question whether Japan should give up nuclear power. This reflects a clear dilemma on the part of the public who are wavering between two worries. They are worried on the one hand about the safety of nuclear power, and on the other about whether Japan is able to move forward if it gives up nuclear power completely. (From the editorial on June 9, 2012 of *Asahi Shimbun*)

Figure 2 below is the poll conducted by the UK in 2011. The result shows that the past 10 years witnessed both a decline and growth in the positive attitude and negative attitude of the public towards nuclear power.

According to the results of the Gallup polls, the Fukushima nuclear accident did not affect the way people in the US felt about nuclear power. The Gallup polls conducted in 1994, 2011 or 2012 indicated that 57% of Americans were in favor of nuclear power. In the past few years the approval rate for nuclear power among Americans ranged from 46% to 62%, with far more men expressing their support than women.

Different Reactions to the Nuclear Accident

With regards to the issue of nuclear power in Taiwan, since the reconstruction of its No. 4 Nuclear Power Plant in Lungmen in 1992, the work on the first reactor at the plant is by and large completed, and the system is undergoing testing.

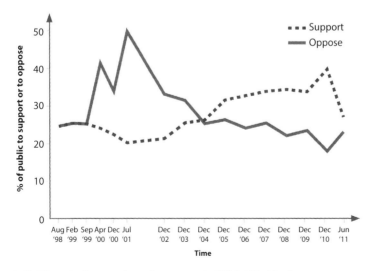

Figure 2 Public attitude towards nuclear power in UK (1998–2011).

As a result of the Fukushima nuclear accident, people's concerns about the safety of nuclear power is on the rise in Taiwan, leading to the call for the suspension of the construction of No. 4 nuclear plant. However, there are also others who hold the opinion that an adequate supply of electricity must be ensured to avoid damaging the economy or affecting people's livelihood. They propose that the construction of the No. 4 nuclear plant be completed according to plan and that necessary measures be taken to guarantee its safe operation.

As a result, the government officially accepted a referendum on the suspension of the construction of No. 4 nuclear plant on February 25, 2013. On March 9, 200,000 people were claimed to have demonstrated against nuclear energy in Taiwan.

In contradistinction to these protests, during March and April 2012, Ma Ying-jeou was re-elected as President of ROC in Taiwan. At the end of his first term and before the renewal of his second term, his government took the initiative to adjust the price of oil and electricity. Chaos reigned in both the ruling party and the opposition. The whole nation raised a terrible clamor. Ma Ying-jeou's government was compelled to take a middle course and defer implementation.

Not long after, the government proposed levying again the stock exchange taxes and those who had benefited before mounted a fierce opposition. In my opinion, even if the price of oil and electricity was adjusted according to the plan, Taiwan would still stand at a fairly low price level as imported energy accounts for 98% of all the energy consumed in Taiwan.

There are various reasons to oppose the adjustment of electricity price. But, the consequence is that the price differentiation will have to be shared out indiscriminately among all the consumers. Yet no one seems to care. Under a misguided notion of egalitarianism, everybody thinks he or she has profited at the expense of others. Therefore, there is no incentive to change an extravagant life-style. They shelve the proposals for improving energy efficiency and environmental protection. Both the general public and the future generation will suffer as a consequence.

The above two cases are indicative of the high levels of awareness among the Japanese of the latest advances in science and technology and the importance they attach to dealing with matters on their own merits.

In addition, the rational manner in which neighboring South Korea reacted to the radiation leaks contrasts starkly with the way panicking people on the mainland and in Hong Kong and Taiwan bought up salt in the belief that it would protect them against thyroid cancer, which can be caused by radiation. Others bought up powdered milk and seafood, thinking that future products from Japan would be contaminated.

Despite intense efforts by the Hong Kong authorities and scientists to present a detailed and balanced analysis of Japan's nuclear disaster, the panic lasted several days, causing a scarcity of salt. These reactions went far beyond the scope of rational thinking, analysis and decision-making.

Interestingly, some local people in Hong Kong suggest that Hong Kong would be safer if Taiwan didn't develop nuclear power, while people in Taiwan say nuclear power plants along the Fujian coastline in mainland China could threaten Taiwan, adding that efforts should be made to oppose nuclear development on the mainland.

So everybody believes everybody else is responsible!

More than ever, we need sustained efforts to make scientific knowledge more accessible to more people. Educating people about technology and encouraging rational thinking are to be applauded, and urged.

4

Aging and Reliability

The damage to the 40-year-old Fukushima Daiichi Nuclear Power Plant caused by the earthquake on March 11, 2011 and the subsequent tsunami is having long-term consequences. However, the nuclear power plant nearest to the epicenter was actually the Onagawa Nuclear Plant in Miyagi prefecture. What does this tell us about reliability and nuclear power plants?

The Onagawa plant is composed of three reactors built in 1984, 1995 and 2002, respectively. It suffered no operational problems or radiation leaks, except for water leakage in some places, following the 9.0-magnitude earthquake and the aftershocks. The three reactors were safe from the 15-meter high waves caused by the tsunami because they are perched high upon a hill. Likewise, the earthquake and the tsunami failed to damage the four comparatively new reactors at Fukushima No. 2 Nuclear Power Plant.

What lessons can we learn from the facts in terms of reliability and safety of nuclear power plants?

Obviously, aging is one of the factors accounting for the disaster of the Fukushima nuclear plant. It had been running for 40 years, reaching the limit of its designed life-span. Equipment may malfunction due to aging as in the case of the nuclear power plant in northern Illinois, US, at the end of January 2012. Aging means there is a shelf-life for everything, from aircraft carriers to space stations and nuclear plants. This means that, in due course, all nuclear plants have to be repaired, replaced, restructured, and then decommissioned when they reach their serviceable-life limit.

Law of Aging

No system is immune to decay, which is why we need to assess reliability in the form of a comprehensive index of operating performance. This index includes three elements: durability, consistency and stability. All equipment, as indicated in Appendix I—old infrastructure, structures with outdated functions, outdated teaching materials—is subject to the laws of aging. Nuclear plants are no exception.

Early on the morning of December 2, 2012 the Sasago Tunnel in Japan, which was built 33 years ago, collapsed without any forewarning or earthquake and 9 people were killed. This disaster was probably caused by the aging of the steel reinforcement in the tunnel.

The transportation infrastructure in the US, such as highways, railways, bridges and airports, was mostly built in the 1950s or 1960s. Highways were mainly built between the 1930s and 1970s. Most of these structures and facilities have aged considerably and sorely need maintenance and rebuilding.

Take America's highway bridges for example. According to official statistics, of the 600,000 or so highway bridges in the US, 27% have been fraught with the problems of structural defect or aging just as the Mississippi Bridge in Minnesota which collapsed in 2007. Barry Le Patner, an American attorney in the building industry, wrote a book entitled *Too Big to Fall: America's Failing Infrastructure and the Way Forward*, in which he depicted the current situation of America's transportation infrastructure and listed the funds needed to maintain or rebuild these facilities.

America's airports and urban infrastructure are faced with the same problem of aging. Here are some typical examples of the grave consequences that aging equipment has incurred in recent years in the US: monumental power failure in the vast north-east region of the nation in 2003; the breaching of the levees in New Orleans in 2005; the collapse of a bridge in Minnesota in 2007; and the collision of the subway trains in

Washington, D.C., in 2009. And of course, the example from Japan: the disastrous incident at the Fukushima Daiichi Nuclear Power Plant in 2011.

For that reason, all aircraft carriers, airplanes, aeronautic and astronautic facilities and nuclear plants have a serviceable-life limit. When they reach that limit, they have to be repaired, replaced, restructured, and then decommissioned with no exception. The nuclear accident at Fukushima definitely has something to do with its commissioned service of 40 years.

Likewise, with the improvement of health and well-being, people in many parts of the world live a longer life today than their predecessors did. The aging phenomenon will become more prominent, resulting in predictable social problems that should be given due attention.

With the vigorous development of its economy, China has been going in for large-scale construction in the past 30 years. As a result, a great number of highways and high-speed railways have been built. Even up-to-standard infrastructure will be confronted with the problem of aging and will be badly in need of maintenance in due course. If the economy happens to stagnate, it will become harder to maintain large-scale aging infrastructure at the same time. We must address this problem from a strategic point of view so as to avert an awful predicament.

The traditional Chinese way of acting and their lack of expertise in reliability design will make things even worse. The infrastructure may be broken into pieces before its aging. Let me cite some recent examples. Since 2007, 18 bridges in China have collapsed for no reason at all, with about 100 people on the death roll. It's worth mentioning that all these are big, costly bridges. One striking example is in Harbin in northern China where a section of the gigantic bridge spanning the city's river collapsed on August 24, 2012. The whole bridge project had cost $120 million.

A great number of reservoirs around the world are facing aging problems now. Has anyone expressed concerns about possible accidents?

When Will Signs of Spring Replace the Old?

The tragedy at the Fukushima Daiichi Nuclear Power Plant must have had something to do with aging: the plant was 40 years old, although poor design, plant location and lack of backup power system are also important factors.

Aging is an issue that should interest all of us. Better health-care and education about health issues mean people in most parts of the world, are living on average much longer than their ancestors. But longer life-spans are leading to social problems that we also need to be aware of, and take precautions against.

There is an old Chinese saying that goes, "Like heavenly bodies going in a robust manner in their own orbits, a gentleman should strive constantly to improve himself". The aged nuclear plant in Japan revealed its fragility under the great waves of the tsunami. The aging problem is like a mountainous tsunami sweeping over our society. Aging in humans is caused by the deterioration of our physiological functions, inhibiting our drive to seek new knowledge and strive for greater creativity. Yet, there are also those in the academic world who show signs of aging long before their time, having lost interest in academic achievements, teaching or research, and are more likely to be washed away by the next generation.

I am reminded of the following lines written by an 11th-century statesman Wang Anshi:

> With firecrackers, the spring wind has ushered in
>   yet another new year in the festive atmosphere;
> With the sunshine penetrating into thousands of households,
>   the new spring signs have replaced the old tokens in the refreshing time.

If you wish to deter aging, you have to revitalize yourself with the new; you must get rid of the decaying and embrace the novel. Constant creativity is a sure guarantee of social progress. Clever old dogs can also learn new tricks, can't they?

5

Transparent Management Guarantees Nuclear Safety

Quantitative analysis, organization and behavior are the three pillars of risk management. The risk management of nuclear power plants has to face the challenges of limited accident data, high safety requirements and a lack of crisis management experience, therefore calling for more vigilance to ensure maximum safety.

The Unit 2 reactor at the Three Mile Island Nuclear Power Plant in the US suffered a core meltdown on March 28, 1979. The cause was faulty machinery and human error, and it led to the world's first serious nuclear accident. Compared to what happened at the Fukushima Daiichi Nuclear Power Plant, the US authorities obviously handled the crisis in a more transparent and efficient way.

During the first two weeks that followed the disaster in Japan, TEPCO tried to suppress the truth and declined assistance from overseas. Today, the rational explanation for this behavior would be that TEPCO had full confidence in its ability to handle the crisis with the technology to hand and, equally important, a huge amount of self-belief. But it wasn't until April 20, a long time after the radiation leaks, that TEPCO embraced advice from France and began filtering contaminated water in the core of the reactor. France's suggestion was highly effective: the contaminated water was diluted to the density of 1/10,000 before being discharged into the sea.

The accident at Three Mile Island could have been far more serious than what later happened at Fukushima. But thanks to prompt decisions by the US authorities and a high degree of transparency, the environmental impact was mitigated.

The research findings of the NRC's 30-year follow-up study suggest that people living near Three Mile Island have not experienced any abnormal health phenomena. The local ecology also appears untainted.

Chernobyl Nuclear Power Plant in Ukraine

One day in the spring of 1986 at midnight, a fire broke out in the 4th reactor at the Chernobyl Nuclear Power Plant in Ukraine (then part of the Soviet Union) which led to a huge explosion. The accident was caused by human error. And in those days, the graphite-moderated reactors used in Ukraine were not protected by any kind of solid containment vessel. The explosion resulted in the most catastrophic nuclear calamity to date.

After the accident, the government sealed the reactor and the debris inside a large concrete sarcophagus to prevent further leaks of radioactive materials. Officials in the then Soviet Union blocked information and refused outside help. Citizens in nearby towns were forced to abandon their homes without any idea of what had happened.

However, when there is a lack of transparent information following a major incident, such as following the nuclear accident in Chernobyl, people give free rein to their imaginations. Many different stories and reports are spread far and wide, causing great fear and confusion.

For example, according to the report released in 2008 by the United Nations Scientific Committee on the Effects of Atomic Radiation (UNSCEAR), the death toll from the nuclear radiation at Chernobyl was 28. But IAEA reported a different number which was 4,000 and that was because it included both early and potential fatalities. The New York Academy of Sciences, on the other hand, published a report by former Soviet Union scientists which gave yet another number for the total fatalities, claiming that almost a million people died of cancer caused by exposure to radiation.

Today, scientific debates rage on regarding the impact of radiation exposure in the Exclusion Zone with one side insisting that chronic exposure to low-level radiation are potentially catastrophic and the other side arguing that plants and animals in the zone are shrugging off the effects of long-term exposure to low levels of radioactivity and are thriving in a fecund wilderness.

It was only 10 years later after the 1986 accident that government officials, scientists and journalists began to visit the exclusion zone surrounding the Chernobyl nuclear plant. The number of visitors grew over the years. Today, the exclusion zone is considered a relatively safe region, and on average, there are approximately 80 visitors who come to visit each day on guided tours organized by travel agencies.

I became one of these visitors of the decommissioned nuclear plant in Chernobyl, which lies some 150 kms from Kiev, the capital of Ukraine, on a freezing day on October 3, 2013. Four reactors were in operation before the nuclear disaster struck on April 26, 1986. The fifth reactor, which was under construction, never got completed. A fact not widely known is that the other three reactors resumed operations about a year after Reactor 4 was destroyed. Reactor 1 was decommissioned in 1991, Reactor 2 in 1996 and Reactor 3 in 2000.

For different reasons, some 2000 elderly people returned to live within the 30-km exclusion zone of the plant a few months after the accident. Over the years, this number has dropped to around 100. One of those still living in the area is Rossana who is in her 90s. Hash, my American-Iranian friend, who is in the nuclear power plant control equipment business, met her on September 27, 2013, five days before my visit to the zone. At the invitation of Rossana, Hash visited her home against the dissuasion of the tour guide.

The background radiation levels range between 0.12 to 1.16 μSv/h within the 30- and 10-km radius exclusion zones. These readings are quite similar to those taken outside the zone. What is more surprising, as Table 6 shows, is that the background radiation levels in some exclusion zones are even much lower than those taken in some buildings in Hong Kong.

There are now a total of 14 checkpoints in the 30-km radius exclusion zone, including those at the 10-km checkpoints and the checkpoint at the small town of Prypyat, where over 10,000 people—nuclear plant workers and their families—once lived. All the buildings in Prypyat, built during the 1970s, are abandoned, including an amusement park where trees, bushes and weeds have grown over the dilapidated buildings. The exclusion zones, vast and sparsely populated, have become wildlife havens, home to many different species of birds and animals, including lynx, beavers, wolves, hares, roe deer, Przewalski's horses, bears and wild boars.

All four reactors, including the destroyed Reactor 4, are closed to outsiders. Fortunately, Mikhail Yastrebenetsky, a Ukrainian scientist, was able to arrange for us to enter the plant to visit some of the facilities and talk with the people there. It was found that while the gamma radiation readings in most places within the shattered nuclear plant range between 0.38 and 3.7 µSv/h, the highest readings are marked at 18 µSv/h. As I walked closer to the 4th reactor, the readings went as high as 19.85 µSv/h. These readings are also shown on the LCD display panel mounted on the wall at the entrance to the plant.

Today some 3,000 engineers and workers are still working at the decommissioned nuclear power plant in shifts, each person working no more than four days a week, or no more than 15 days a month. They test the water and the soil for radioactive elements and work to contain the contamination. They are also constructing a giant hangar-like metal shelter for the destroyed 4th reactor. The construction project is supported by international funding managed by the European Bank for Reconstruction and Development, and it involved professionals from France and other countries in addition to Ukraine. The shelter is expected to be completed in 2015 and will be placed permanently over the sarcophagus to avoid radiation leaks.

TEPCO's Crisis Management

In Japan, it's another story: TEPCO failed to take effective action quickly enough, still claiming in mid-April of 2011 that the Unit 5 and Unit 6 reactors should be put back into operation as soon as possible. The company was even proposing in May that new nuclear reactors, Unit 7 and Unit 8, be set up on the site of the Fukushima power plant. Even the mildly cynical might say that TEPCO's motivations were commercial rather than in the public interest.

According to a report on *The Wall Street Journal* on May 18, TEPCO engineers had underestimated the consequence of the nuclear accident, and had overestimated their ability to repair the damage, an error that resulted in further damage and eventually categorization as a level-7 accident on the International Nuclear and Radiological Event Scale, the highest grade possible for nuclear accidents. Consequently, the public lost trust in TEPCO.

A series of events following the accident unfortunately proved right a point I made to *Ming Pao Daily* the day after the Fukushima nuclear accident. According to the declassified minutes of the meetings of the Nuclear Disaster Countermeasure Headquarters, one independent expert warned at a meeting held four hours after the terrible tsunami had hit the Fukushima nuclear plant that the reactor might go into meltdown if the temperature at the core should rise over the next eight hours. However, the Japanese government and TEPCO did not admit until May that there had been a meltdown in three of the six reactors in the nuclear power plant.

I had predicted at the time that a leak of the coolant water would certainly pollute the nearby soil environment. Unfortunately, my prediction was confirmed by the report released by TEPCO on April 6, 2013.

On March 14, ships from the US Navy that had been assisting rescue efforts withdrew to 80 km off the Japanese coast, shocking the world into thinking that the crisis was reaching catastrophic levels. Some public opinion held that residents within 80 km of the plant should withdraw as well.

I guessed the evacuation might have something to do with the cooling water. My hunch turned out to be true. On April 5, TEPCO declared it planned to discharge 11,500 tons of low-level radioactive water into the sea, so for safety's sake, the US Navy had been advised to sail away.

Again, according to a report on July 12, 2013 in the Japanese Yomiuri Shimbun, TEPCO acknowledged that radioactive substances were leaked from Fukushima Daiichi nuclear plant after the earthquake on March 11, 2011.

Recent Events at Fukushima Daiichi Nuclear Power Plant

Due to the fuel-rod meltdowns at three nuclear reactors, Japan's Nuclear Regulation Authority (NRA) concluded in June 2013 that three out of the seven storage pits are leaking, compounding clean-up difficulties. This adds additional challenges to move all contaminated water containing tons of high level of radioactive cesium-137 to the above ground tanks.

On my trip to Fukushima Daiichi Nuclear Power Plant on July 2, 2013, I noticed some 100 steel tanks that TEPCO has built across the plant complex to store radioactive water coming from the reactors as well as underground water running into the reactors and turbine basements. However, as of August 23, 2013, 300 tons of contaminated water was confirmed to have leaked from at least one of the tanks, possibly through a seam.

The leak is the worst since 2012 involving tanks of the same design at the wrecked Daiichi plant, raising concerns that contaminated water could begin leaking from storage tanks one after another. That could create extensive soil contamination and was a blow to plans to release untainted underground water into the sea as part of the efforts to reduce the amount of radioactive water.

The leak has weakened confidence in the quality and reliability of these tanks that are crucial for storing water to be pumped into the broken reactors to keep the radioactive fuel cool. It also deepens further concerns of the public on the crisis management of the Daiichi Nuclear Power Plant.

Since 2011, the Fukushima Daiichi plant has been leaking hundreds of tons of contaminated water into the sea which heightened the sense of rising crisis and also caused panics in the neighboring countries and areas. More than two years after the accident, TEPCO is still unable to control the discharge of contaminated water from its wrecked Fukushima plant due to a series of mistakes made by the management to tackle the problems and often to hide the problems.

Instead of leaving this up to TEPCO, the Japanese government decided on September 3, 2013 to take over the task and planned to spend $470 million on building a subterranean ice wall and on upgrading the water treatment facilities.

The concept of underground ice wall has once been used to isolate radioactive waste at a site of Oak Ridge National Laboratory (ORNL), Oak Ridge, Tennessee, according to *MIT Technology Review*. It will take two years for the Japanese government to build the unprecedented 1.4 km wall to surround the reactor buildings and their adjacent facilities at the Fukushima plant.

At the same time, Unit 3 and Unit 4 reactors at the Oi Nuclear Power Plant, the only nuclear power reactors in operation as of September 2013, are shut down for routine inspection (see Chapter 7, Nuclear power and nuclear industry in Japan).

Rules-based Management versus People-based Management

As far as management style is concerned, the Japanese are accustomed to rigid systems. Of course, there is much to commend about their approach to work: they carry out tasks in a thorough manner; they commit themselves to their professions and duties; and there were even the "Nuclear Warriors" who exposed themselves to radiation while battling the fallout from the stricken reactors.

However, the Japanese involved in the disaster operation could do nothing before their reports had been approved by the highest-level authority, and they could not act until they had written instructions for handling all contingencies. Under such circumstances precious time was lost. In contrast, the US system is more flexible. The person in charge is granted executive powers, while managers at different levels usually have a higher degree of autonomy and respect.

If I can make a comparison, I would say that the Japanese prioritize the "rule-based management", whereas the Americans value "people-based management". A lesson learnt from handling the Three Mile Island nuclear accident reveals that "people-based management" is often more efficient in emergencies, as long as the right person is in charge.

Two years after the Fukushima nuclear accident we are still analyzing how this devastating tragedy has impacted people's mind. A sense of insecurity, which seems disproportionate relative to the known facts, does not arise solely from a lack of scientific knowledge. In my opinion, the method and attitude used to address the crisis have made a significant negative contribution to the current crisis management effort. Indeed: the way TEPCO dealt with the nuclear accident should serve as a negative case example for Executive Master of Business Administration (EMBA) programs.

6

The Need for Quality Control

The concept of modern quality control originated in the US at the beginning of the 20th century through companies such as AT&T. After World War II, the concept filtered into Japan where it was further developed into the notion of "total quality control," comprising market investigation, development, design and post-sale service. In the last 40 years of the previous century, the Japanese semiconductor and computer products business and the automobile industry succeeded thanks in general to this concept. In recent years, it has found its way into the service sector.

The frequently used word "accountability" is a concrete example of quality control practice.

When Toyota introduced the "quality control circle" concept, the purpose was to encourage its employees to improve quality control as a team by means of group discussions. Later, the concept of "quality control circle" spread from the manufacturing to the service industry and other large-scale professional organizations.

At present, the concept of "quality control circle" has been replaced by that of "accountability". This new concept means that every member of a group should have "team spirit". Each is accountable at different levels and each member is required to do his/her bit in his/her position. The essence of "team spirit" lies in the accountability of every team member. That means every team member is responsible for improving their team's performance. The minimum requirement is to "go on tolling the bell as long as one is a monk".

"Accountability" applies to all levels and all members of a team instead of a specific stratum. "Accountability" also applies to every team in society. Otherwise, won't the whole world be thrown into disorder?

An Ancient Doctor's Comment on Quality Control

Quality control is no trivial matter. Rather than looking for remedies after quality-related accidents have incurred great losses on us, we should remove potential perils before an accident occurs. This reminds me of a classic story which, I think, best illustrates the importance of quality control. The story has to do with a dialogue between Duke Wen of the Wei State and Bian Que, a famous doctor from my ancestral hometown Cangzhou in Hebei province.

Duke Wen once asked Bian Que, "You and your two elder brothers all practice medicine. Which one of you is the best doctor?"

Bian Que said, "My eldest brother is the best one. The other brother is second to him, but still better than me."

Duke Wen asked again, "Then why are you best known?"

Bian Que answered, "My eldest brother cures his patients before they develop an illness. His patients have no idea that he has helped them remove the cause of illness. That's why his name never gets spread around. Only we, his family members, know the depth of his skill. My second brother cures his patients before their condition worsens, and people think he can only treat some minor illnesses. That's why he is only famous in our village. As for me, I always cure patients who are very sick. When people see me perform phlebotomy and major surgery, they think I'm capable of treating all kinds of difficult and complicated cases and saving them from the brink of death. That's why they think I have the best skill and I became famous throughout the kingdom."

This story tells us that a good doctor is one who is skillful at preventing diseases. In a succinct way, it illustrates the truth that prevention is better than cure. It is better to prevent an accident from happening or control it at an initial stage than to repair what has been damaged later. Overall quality control means the quality control of personnel, systems, procedures, products and services. The control and improvement of quality require concrete assessment standards.

This quality control concept, extended to quality assurance and then quality design, has been widely practiced in advanced industrialized countries. In the past dozen years, it has become very popular in South Korea, and has been extended from manufacturing and service sectors to the government, public utilities, labor, the film industry, hospitals and the higher education sector. The flourishing of science and technology in South Korea at the end of last century was attributable not only to their successful innovations, but also to quality assurance. Reliable evaluation depends on objective assessment standards.

Appraisal and rankings are in fact practiced in all businesses: *Consumer Reports* advise people on what cars to buy; commodities are subject to best brand selection exercises; and hotels and tourist destinations are rated according to a star system. In the area of higher education, there also exist various kinds of evaluation mechanisms. Many students and their parents fill in their admission applications after taking into consideration the ranking of universities. Many big enterprises also make their decisions for granting applicants' interviews and setting their salaries on the basis of the ranking of the applicants' alma maters.

Even Heaven would perhaps also be subject to quality control.

Performance of Power Plants

In the nuclear power industry, the performance of a plant is measured by its capacity factor. The net capacity factor refers to the ratio of the average load and the capacity in the electrical device at a given time. A low-capacity factor might be due to several possibilities: low efficiency (unstable wind-powered electricity), time-consuming maintenance (various power plants), high frequency of maintenance (power plant accidents, natural disasters and human-induced accidents), or poor reliability (brands or operating errors).

According to a *Nucleonics Week* report on June 9, 2011, the average capacity factor of the 54 operating nuclear reactors in Japan before the Fukushima accident was much lower than the average—ranked 26th among the 443 nuclear reactors in over 30 countries and regions. Indeed, measured by quality, nuclear power plants in Japan have been exposed to dangerous factors, and quality issues have been leveled at the Fukushima nuclear plant in the past.

Nuclear power is closely watched and has always been subjected to rigorous quality control across the world. In addition to customer supervision and academic consultation, governments usually set up independent supervisory bodies to ensure quality and safety, such as the NRC in the US, and the Nuclear Safety Commission (NSC) in Japan. On September 19, 2012, NSC was reorganized to form the Nuclear Regulation Authority (NRA). In addition, the Japan Nuclear Energy Safety Organization (JNES) executes the assessment, inspection and approval of nuclear power plants. JNES was set up in 2003 as an expert body to inspect nuclear installations and to undertake safety analyses. The Japanese cabinet on October 25, 2013 approved the merger of JNES with NRA.

In-breeding Encroaches upon Quality Control in Nuclear Plants

Under ideal circumstances, maintaining co-operation and checks and balances among the power plants, JNES and academic circles in Japan should function as an effective means of ensuring safety. However, "inbreeding" is a wide-spread phenomenon in Japan. Complex socio-cultural and economic ties in Japan are obstructive, narrowing the scope of broad knowledge dissemination for the short-term, while longer-term negative effects can be so serious that safety might be undermined by self-serving interests and personal ties and loyalties, derived from shared backgrounds among top administrators, in particular through the old school-tie network. For nuclear power plants that call for the highest degree of safety, this is a deeply worrying phenomenon. After all, when we look at the Fukushima nuclear accident, safety measures failed to function, which was a source of great criticism.

In early July 2012, the fact-finding mission composed of 13 experts appointed by the Japanese government released a detailed report that lays bare the fact that the administrative body of TEPCO in Japan covered up the potential risk to the nuclear power plant. It also reveals that the power company demanded that the Japanese government de-emphasize the possibility that Fukushima might be struck by serious tsunamis. (Refer to Postscript) In my opinion, this report came too late.

Quality control has led to success and failure in Japan. If not for the Fukushima nuclear accident induced by the earthquake and tsunami, no one could have believed that quality would be an Achilles' heel for the Japanese, where quality control is deemed a societal norm. Similarly, in mainland China, Hong Kong, Taiwan and Macau, "in-breeding" in recruitment process is no less prevalent, which is bound to hinder progress and threaten safety in these societies over the long-term.

Japan's new Nuclear Regulation Authority (NRA) was established after the Fukushima accident in response to widespread distrust of the previous regulatory agencies. The responsibility of NRA is to determine and regulate the location of Japan's nuclear power plants to ensure that they are built on safe grounds.

7

Don't Let Gossip Affect the Safe Operation of Nuclear Power

Since the Fukushima nuclear accident, quite a few unfounded rumors have come into circulation: the "Fukushima 50", i.e. the alias given by the media to 50 TEPCO employees who remained at the crippled Fukushima Daiichi Nuclear Power Plant after others had evacuated, would die within two weeks; the incident was like an activated atomic bomb; or the nuclear accident has led to the deaths of more than 9 million people in the world.

One even goes so far as to say that radioactive materials released from the damaged Fukushima plant will be carried underground to Beijing and Xi'an in China, and even to the west coast of the US; while another one maintains that as the percentage of nuclear power in China's energy structure is low there is no need to increase it; still another claims that it is one of the tricks of the rich countries to encourage the third world countries to use nuclear power, and so on. These soothsayers are so full of conviction that they are ready to swear their lives by them.

The persistence of these baseless rumors reminds me of Su Shi's poem "Shadows of Flowers":

> *The terrace is covered with layer upon layer of shadows of flowers,*
> *Boy servants are told time and again to sweep them away to no avail.*

The shadows seem to be cleared away by the sun when it goes down,
They reappear again on the terrace by the light of the bright moon.

As it is hard to clear away all the dead autumn leaves, the Fukushima nuclear accident, being a natural disaster aggravated by human error, will certainly impact people and the environment. In reality, they are but baseless and unnecessarily pessimistic rumors not worth refuting.

Self-defeating Hearsay

Here is a modern-day story in the style of *Strange Tales from a Lonely Studio*. On her way back home after she had finished her night shift, a woman found herself being tailed.

An idea came to her. She turned into a graveyard, and said loudly in front of a tombstone, "Open the door, Dad. I'm home." The man who was shadowing her was so terrified that he took to his heels. Overjoyed, the woman was about to leave when she heard a voice coming from the grave: "My girl, you forgot to take the key with you again." The woman fainted from fear, not realizing that by coincidence, a grave-robber was also out working that night.

More than one year after the Fukushima accident, six workers who had joined in the rescue work of the plant died of illness or injuries, "but none of the deaths were linked to radiation," stated a press release on May 23, 2012 from UNSCEAR based in Vienna. So in our search for the truth, we must focus on the issue at hand, and avoid groundless speculations, or getting panicked for no reason in the process.

It is understandable that nuclear power can cause alarm and confusion among the general public. Most people are unfamiliar with the science and technology that go into the nuclear power industry. For example, the day after the accident, the mass media in Hong Kong and Taiwan predicted that the "Fukushima 50" would die within two weeks. This forecast turned out to be incorrect.

Some people are afraid of ghosts, but they are rarely frightened to death by ghosts! In order to seek the truth, we should base our judgment on the facts and avoid speculation and rumor. Otherwise, instead of being frightened to death by the ghost, you'll be scared to death by yourself.

Nuclear power is an innovation in the peaceful use of atomic energy; it has its own value when other reliable alternative energies are not available. To be sure, nuclear safety must not be ignored since the consequences of a serious nuclear radiation leak even in foreign countries like Japan could be painful beyond imagination to the public elsewhere even though far away, especially when one considers the psychological stress on the public and the accompanying social cost.

Three Elements in the Safety of Nuclear Power

Nuclear power safety is contingent upon three major elements: equipment, personnel and operation management. As far as the first element is concerned, most of the mainstream nuclear power plants in operation or under construction are equipped with Generation II reactors or an improved version, with very high levels of safety. The Generation III reactors and the Generation IV reactors being developed now will further improve the safety and functionality of the plants.

The other two key elements in nuclear safety are personnel and operation management. The malfunctioning of these two elements was the primary cause of the previous nuclear accidents. Both accidents at Chernobyl and Three Mile Island happened at midnight and were caused by human error. In the design of any new nuclear equipment, therefore, one has to take into consideration the concept of fault tolerance to lessen the negative consequences of human error.

The key to the safety of a nuclear plant is to optimize management once the plant is put into operation. As all equipment and systems are subject to wear and tear, including nuclear reactors with strict safeguards, we must remain vigilant and follow the established procedures of operation and emergency-management, allowing no room for negligence.

The US claimed that the equipment of the country's nuclear power plants is good enough to cope with natural disasters. But a report in the *New York Times* pointed out in May 2011, however, after an inspection on 104 reactors across the nation, that the NRC had discovered that factors such as those resulting in the Fukushima nuclear accident had not been taken into consideration in the country's existing disaster response programs. If there is any neglect in the normal operation management of the plants, there is no guarantee that an accident will not give rise to a disaster.

Warnings about flawed maintenance work were repeatedly issued against the Fukushima nuclear plant before the accident. As it turned out, the backup cables in the plant were swamped by the tsunami that followed the earthquake and the backup DC generator proved to be of no use — all these reflect precisely that the operation management of the plant was as much flawed as the initial design.

We cannot rely exclusively on private corporations to ensure the safety of nuclear power since, with considerations different from those of the public, they will prioritize cost and profit in the implementation of safety measures.

In the Fukushima nuclear accident, for example, TEPCO had to submit its report upward through the hierarchy in accordance with relevant government regulations; and the company was reluctant to abandon the crippled plant due to commercial interests, and the optimal time of the rescue operation was thus allowed to lapse. The nuclear safety crisis in Japan was no doubt triggered by natural disasters, but it is also true that TEPCO's management did not respond to the warnings about their flawed maintenance by taking actions that would have consolidated and protected the backup power generators before the crisis; and failed to take prompt reinforcement measures during the crisis. The backup power would have worked if it had been located at an elevation higher than the tsunami.

To ensure safety, therefore, nuclear power operation must be put under the joint supervision of the government and the public.

Nuclear Power and Nuclear Energy Industry in Japan

Before the Fukushima accident, nuclear power accounted for about one-third of Japan's power consumption, and the country planned to increase it to 50%. After the accident, however, many people argued that Japan should abandon nuclear power. Japan's nuclear plants began their commercial operations in 1966, and by 2011 there were 54 nuclear reactors in total.

After the incident, local governments in Japan did not authorize the nuclear reactors that had been shut down for periodic safety inspections to restart. By March 27, 2012, Unit 3 at the Tomari Nuclear Power Plant in Hokkaido was the last of the nation's 54 commercial reactors still in operation.

According to a report from a Japanese source, Units 1 to 4 in TEPCO's Fukushima plant officially went out of commission on April 19, 2012, and the number of the nation's commercial reactors went down to 50 starting from April 20. As the summer electricity consumption peak season approached and electricity supply was in short supply throughout the country, a four-member team, which was led by Prime Minister Yoshihiko Noda, aiming at the restart of nuclear reactors, concluded that the Unit 3 and Unit 4 reactors at Oi Nuclear Power Plant in Fukui prefecture met safety standards, and hoped the local government would agree to restart them.

For a time Japan's METI failed to persuade the local governments to consent to the restart of reactors. After the Tomari Nuclear Power Plant suspended the operation of its Unit 3 reactor on May 5, 2012, Japan officially entered the stage of no nuclear power. The consequences are an increase in the electricity price, failure to achieve its CO_2 emission goal for 2011 due to switching back to fossil fuels, and a report of a trade deficit for the first time in 30 years, a situation that grew worse in 2012 and 2013.

As was expected, due to the pressure from power shortages and the approach of the hot summer, Japan's government announced in early June the restart of the Oi Nuclear Power Plant at Fukui prefecture, reversing its decision after less than two months, and rejoining the world of nuclear power. On March 9, 2013, two days before the second anniversary of the Fukushima nuclear accident, 15,000 people staged a demonstration against nuclear power in Tokyo.

Japan prepared to restart nuclear plants two years after the tsunami smashed into the Fukushima Daiichi power plant following which only two of Japan's 50 functional reactors had been in operation. Utilities had been forced to run coal-, gas-, and oil-fired plants to make up for nuclear power, which once supplied 30 percent of Japan's electricity. But in the elections to the House of Councillors held in July 2013, the Liberal Democratic Party of Japan, a pro-nuclear party, succeeded in consolidating power just as utilities were submitting applications to restart reactors.

As of mid-July 2013, applications have already been submitted to restart 14 of the 50 nuclear power reactors idled due to the loss of confidence in Japan's nuclear safety and risk management during and after the Fukushima accident. The speed for NRA to process the applications under its new safety rules is dependent on its manpower. If the staffing level is right adjusted, some 28 units are expected to put back in operation by April 2015, said the Institute of Energy Economics, Japan (IEEJ). The country will then save more than $8 billion in fossil fuel imports.

While Japan is cutting down on the domestic use of nuclear power, it is actively seeking to export its nuclear power technology, including negotiations with the government of Vietnam in October 2011 for a $13 billion project to build two reactors there. In addition, Japanese manufacturers are bidding on, or working under contracts for, nuclear plant projects in other countries and regions including China.

Although Japan experienced the Fukushima nuclear catastrophe, it has taken active measures to improve nuclear safety and succeeded in exporting nuclear power technology to Turkey in 2013.

Before the Fukushima accident, the Japanese government listed the expansion of trade in the export of nuclear power technology as one of its core growth strategies, with a target to export two reactors each year, totaling 50 units by 2030.

South Korea: An Emerging Power of the 21st Century

The Republic of Korea (South Korea) has become a manufacturing powerhouse, producing high-quality electronic products and automobiles. The country is also closely connected to the rest of the world through higher education, the cultural and creative industries, and its democratic political system. But a more note-worthy aspect, which many people might not realize, is its attention to energy and the environment. South Korea has mapped out strategies to become a world exporter of energy and environmental protection by developing its own nuclear energy.

In the 1980s, when the plan to build the No. 4 Nuclear Power Plant in Lungmen based on the Combustion Engineering technologies was suspended in Taiwan's Chiang Ching-kuo era of 1980s, South Korea seized the opportunity to take over from the US to develop nuclear technology at full speed. It is now poised to surpass Japan to share the limelight with the world's other leading countries in nuclear energy production.

South Korea, a latecomer in the nuclear power industry, has drawn up a strategic plan for the development of nuclear power and is stepping up its effort in enhancing its capabilities in nuclear research and developing the manufacturing sector of the midstream and upstream nuclear power plants.

Today, KSNP+ (Korean Standard Nuclear Power Plant) with its modified design of OPR-1000 (Optimized Power Reactor) has become South Korea's hallmark technological achievement following such world-famous Korean brands as Samsung and Hyundai. South Korea's strategic plan is aimed at turning the nuclear-power-oriented manufacturing and service sectors into the most profitable industry, second only to the automobile, semiconductor and shipbuilding industries.

South Korea once estimated that 59% of electricity consumed in the country will be supplied by its nuclear power plants by 2030, and plans to export 80 nuclear reactors by 2030, thus turning itself into the world's third largest exporter of nuclear power equipment after the US and France or Russia. In addition, in order to meet the demands of the global market for nuclear power and address the heavy reliance of various countries on energy sources, South Korea is striving to claim its share in the world market in technical assistance to the operation of nuclear power plants, their maintenance and after-sale service.

China: Nuclear Power House in the 21st Century

China's first nuclear power plant, constructed in Qinshan, Zhejiang province, was put into operation in 1991. China's subsequent nuclear development was however modest until 2005. Increasing concerns about severe air pollution and abnormal weather patterns, rapid economic growth and severe power shortage have driven China to look for alternative energy supply, other than hydroelectric power, to reduce its heavy reliance on coal burning for electricity.

Due to a widespread power shortage from 2002 to 2004, former Premier Wen Jiapao announced in March 2005 the need to adjust China's energy mix and to vigorously develop nuclear power. This led to the State Council's approval in March 2006 of the "Medium and Long Term Nuclear Power Development Plan" (2005–2020), and white paper "China's Energy Conditions and Policies" which listed nuclear energy as an indispensable source for the country. Since then, China's nuclear power has developed at a pace which is unprecedented elsewhere in the world.

As of 2013, China has 17 nuclear reactors in operation with a total capacity of 14.69 gigawatts (gW). There are 28 more under construction with a total capacity of 30.57 gW which is about 40% of the combined capacity of all the nuclear reactors under construction in the world. The target indicated by China's National Development and Reform Commission is to raise the percentage of China's electricity produced by nuclear power from currently less than 2% (Figure 1) to 6%, a total capacity of 86 gW, by 2020. A total of 105 nuclear power reactors (Figure 5) are expected in operation by 2035.

Regarding nuclear technologies, China has imported them from France, Canada, US, Russia and others and is actively seeking to maximize self-reliance in the design and manufacturing of nuclear power plants through international cooperation and technology transfers. Advanced pressurized water reactors such as ACPR1000, AP1000 and CAP1400 are some of the examples of China's efforts in developing nuclear technologies, along with the pebble bed reactors. Fast breeder reactors are being planned to contribute to adding more nuclear capacity. China is also involved in the development of nuclear fusion reactors as well as research and development into the thorium fuel cycle as a potential alternative means of nuclear fission.

All these efforts pave ways for China to establish its long-term plans for future nuclear reactors which include a total capacity of 150 gW by 2030 and 350 gW by 2050. To complete the fuel cycle management of its nuclear use, China is drawing up clearly defined workable timetables for the final treatment of high-level nuclear waste in the deep underground rocks in Gansu province.

Following the Fukushima accident, China announced on March 16, 2011 that all nuclear project approvals were frozen, approvals for construction of new plants, including those in the preliminary stage of development, were temporarily suspended, and comprehensive safety checks of existing reactors were ordered. Although the Chinese government has since 2012 indicated to continue its overall nuclear energy strategy, many believed that safety and economic development related risks could cause a further review in favor of a full spectrum of rainbow energy options.

As the fastest expansion of nuclear power in the world and in light of the repeatedly mishandling of the polluted water at the Fukushima Daiichi Nuclear Power Plants site by TEPCO during May and August 2013, the public in China and South Korea will likely pay extra attention to the safety and reliability issues of the operations of their nuclear power plants.

Furthermore, rapid nuclear expansion in China and elsewhere in the world may lead to a shortfall of nuclear fuel and qualified nuclear and reliability engineers. Significant human resources training and strengthened regulatory and safety enforcement regimes are being developed in China to support its aggressive peaceful use of nuclear energy program.

Taiwan: Confounded by the Nuclear Power Issue

Taiwan has a 30-year history of nuclear electricity and a 50-year history of nuclear energy, with three nuclear power plants currently in operation and a fourth at Lungmen which has been under construction for the last 20-odd years, and is still not yet completed.

After the Fukushima nuclear accident, almost the entire Taiwan community voiced fierce criticism of nuclear power once again, claiming that the No. 4 Nuclear Power Plant in Lungmen was the result of assembled parts from the US, Japan and Taiwan behind close-doors, and concluding that the plant would be hazardous and unreliable.

It is totally groundless to assume that the No. 4 Nuclear Power Plant in Lungmen is not safe simply because it is built with assembled parts. A careful observation of our surrounding facilities and daily necessities tells us that almost everything today is assembled, from highly sophisticated instruments, domestic electric appliances to electromechanical products for military and civilian uses. For instance, aren't all the cars, planes, computers, cell phones, TV sets, food, and even the cast in Ang Lee's Oscar-winning films assembled through cross-industry, cross-trade, cross-species or international cooperation?

Take people relatively advanced in age for example. Often on their bodies, we can find artificial skin, transplanted organs, and soft and hard biochemical devices and electromechanical appliances. These gadgets are also assembled or combined products, regardless of whether they are intended for health or cosmetic purposes. These assembled pieces, after being planted into human bodies, run smoothly, performing the double function of improving quality of life and promoting longevity. They help to cultivate a life of ease and comfort, and it is hard to imagine people who do not someday have one or more external devices assembled or implanted in their bodies to guarantee a more pleasant life.

We are living in a world full of assembled objects and composites. It is perfectly all right so long as the different pieces are combined in a proper way; that is, the different elements agree in size, are complementary to each other, and appropriately employed. They can be adroitly suited to various circumstances because of their adaptability and versatility. I can see nothing wrong with them. In history, there were many corporations who obstinately refused to form conglomerates or establish alliances, such as AT&T, Trans World Airline (TWA), and Wang Computer. As a consequence, these companies lost their competitive edge, and eventually vanished mainly because their products were not catching up with market demands.

The No. 4 Nuclear Power Plant in Lungmen has experienced several ups and downs since the drawing-room stage, going through a series of interruptions and resumptions of work, and contract disputes. As a result, the construction has been chaotic, projects had to be outsourced, testing results were found to have deviated from standards, and so forth, although in the initial stage, deviation from the norm is usually not uncommon (see Appendix I). In the case of the No. 4 Nuclear Power Plant in Lungmen, due to political and social factors, the design and construction have been delayed or hindered for as long as 20 years.

Top priority should now be given to high quality control for implementation at all levels, carrying out stringent project inspection, and applying reinforcement measures to both the software and hardware projects to ensure that the reactors will eventually meet reliability criteria. If these steps are followed, the problems that have popped up during the early stages of the construction can be solved.

Speaking from Evidence

The essence of reliability assessment lies in speaking from evidence. Building on an experience which is much shallower than that of Taiwan, South Korea is now well on its way into the international market, notching up achievements that could place them one day on a par with nuclear power giants such as the US and France. In the meantime, Taiwan, alas, is still gripped in a bitter dispute over the issue whether or not it should continue efforts to build a long overdue nuclear power plant. Precious time has been wasted, one way or the other.

Based on the past 30 years' safety record in nuclear power plants operation in Taiwan, we can find no firm basis for making any hasty conclusion or judgment without a thorough investigation in the case of the No. 4 Nuclear Power Plant in Lungmen. If we are concerned with nuclear waste, has Taiwan society at large, other than the Institute of Nuclear Energy Research, ever made any serious efforts to advance their knowledge of, or improve their solutions to handle or to store, nuclear waste?

Besides, if we find nuclear energy really unacceptable, have we come up with a reliable, feasible energy and environmental protection strategy that is sufficiently convincing and practicable in Taiwan? Why don't we simply abandon the plant? Have we done any research to come up with alternative sustainable energy and environmental protection products that are feasible for Taiwan?

Just as a beautiful autumn flower refuses to complain about the untimely spring wind, a sensible person will not complain about his lack of luck. At the moment, South Korea attaches immense value to going full steam ahead in the pursuit of excellence in various fields in the globalized 21st century while Taiwan, overwhelmed by fear, is still hesitating, crippled by indecisions that greatly hinder advancement. The dispute over nuclear power reflects the social realities in Taiwan today. It continues to lag behind the times in terms of internationalization and higher education, and in such key industries as automobile manufacturing as well.

Why is it that a beautiful autumn flower refuses to complain about the untimely spring wind whereas there are people who make a big fuss about the heated summer?

The Unsettling Element in Discussions about Nuclear Power

At first glance, nuclear power safety might seem to be just *one* issue, but in fact there are *two* implications.

The first implication is reliability, which is related to the probability of accidents happening in the plant; the lower the probability, the higher the reliability. The second implication is security, similar to what is generally called "safety" and "assurance." In other words, in the event of an accident in the nuclear power plant, can the residents living nearby be evacuated from the site in a timely and safe manner to avoid exposure to levels of nuclear radiation which might increase future incidences of cancer or other diseases?

Every time I give a speech, the question I put to people in the audience most frequently is how many people, do they think, died from nuclear radiation following the Fukushima nuclear accident on March 11, 2011.

To my surprise, more often than not, their ready answer is the same: the death toll reached more than 10,000. The day after the Fukushima nuclear accident, the media in Hong Kong and Taiwan reported that the 50 rescue workers would certainly die within two weeks, but their estimation proved to be untrue.

Some correspondents have asked me why they have not read any reports about it if it were true that nobody had died during this nuclear accident. I replied, "The fact that there is no such report is positive proof that nothing of that kind ever occurred. Otherwise, how could those reporters ignore such a great news story?"

In view of the two factors of reliability and security mentioned above, researchers have spent years tracking, observing and recording nuclear power accidents such as the Three Mile Island incident in the US and the Chernobyl accident in the former Soviet Union, and used the results to calculate the reliability of nuclear power. However, as most people know little about nuclear power, once an accident happens, the consequences can be easily blown out of proportion.

The nuclear power plant in Fukushima is designed to withstand 7.0-magnitude earthquakes and waves up to four meters high. It was not affected in any significant way by the 9.0-magnitude earthquake in 2011, and could have resumed operation anytime had the devastating tsunami surges not followed.

Many people find it hard to escape from the haunting shadow cast upon them by nuclear power. In other words, whether it is safe or not, they just feel uneasy about it. Feelings of insecurity are but human nature. Infants and children are afraid of the dark even when they know that they will sleep through the night and wake up safe and sound in the morning, but the fear of darkness still weighs heavily on their minds. With some, it may stay with them until they die.

As a source of electricity, nuclear power will have to face two safety issues in its operation: one is the site of the plant, and the other is the disposal methods of nuclear waste.

Many people can accept nuclear power on the condition that the plants and the waste repositories are far from their own homes. Such a state of mind is reflected in a saying in Taiwan, "The children of others won't all die", and is similar to what is called Nimby ("Not in My Back Yard") in English, although some polls conducted by Bisconti Research Inc. and Quest Global Research Group in the US indicated that a high percentage of people living within 20 km of nuclear reactors favored the use of nuclear energy. There are also people who manipulate the public's sense of insecurity, deliberately exaggerating the risks of nuclear power. Under the pretext of responding to public opinion and showing concern for environmental protection, their primary goal is to build up their political capital.

Feelings of insecurity and the self-serving and myopic practices of politicians are part of the reality that draws the public into an unwilling frenzied dance with them. We must face up to these issues. It is understandable that public concern about the safety of nuclear power could lead to collective anxieties, but we must, with a scientific mindset, let the evidence speak for itself, or we shall continue to be troubled and behave like the man in the fable who worries that the sky will fall on his head one day. Only then can we avoid getting panicked unnecessarily.

If our knowledge proves that nuclear power is safe, we can live on with peace in our mind, and enjoy the utmost benefit that nuclear power can afford us. Otherwise, we will continue to be plagued by such unfounded worries.

Or is it human nature to harbor such worries?

Part Two

Environmental Protection, Occupational Safety and Innovation — A Spectrum of Energies

8

A Spectrum of Energy Sources

In my last year at National Tsing Hua University, each student had to write a research report. The research program I chose was neutron activation analysis with a focus on chemical pollution in the streams and rivers of Taiwan. Neutron Activation Analysis is a process using the method of irradiation through neutrons in nuclear reactors to examine the energy spectrum of γ rays in samples containing tiny amounts—at the ppm level or less—of poisonous elements, such as mercury and arsenic, for example. This process is used to check the safety of food and drinking water.

Neutron Activation Analysis is just one of the many ways nuclear power is used to protect health and support the economy. For example, Japan experienced heavily polluted rice and drinking water in the 1960s due to excessive use of chemical fertilizer. Similar problems have occurred in mainland China and the mid-west agricultural states in the US.

As a staple food, rice must be safe to eat, and be provided for on a regular basis and within the purchasing power of the average consumer. Even a tiny amount of toxic elements is harmful, and were not always detected by tests in the past. Today, nuclear power helps guarantee safety through the use of Neutron Activation Analysis.

A Brief History of Energies

Ever since the discovery of fire, firewood has occupied a very important position in the human history of energy. Just as Wang Yuxiao described 1,000 years ago in his poem "The Festival of Pure Brightness":

> I passed the Festival of Pure Brightness without the
>   enjoyment of flowers or wine,
> I feel lonely and bored like a monk practicing
>   Buddhism in a secluded temple.
> Yesterday my neighbor rekindled a new fire after
>   the Cold Food Festival,
> The light glowed through the window to serve
>   as my reading lamp.

Although the poor scholar lived a lonely and boring life, he made use of his neighbor's light and concentrated on his reading. He was so content with his lot that he finally achieved a great deal. From this example we can see the importance of the role of firewood in our lives. In the early period of human history, in addition to firewood, renewable energies included manpower, animal power, water power and wind power.

The Industrial Revolution in western Europe promoted the mining of coal on a large scale. In 1860 coal accounted for 24% of all the energy sources consumed in the world. The ratio rose to 62% in 1920. In the 1970s electric power replaced steam engines and promoted the rapid development of the electric industry. In 1965 oil replaced coal as the highest-consumed energy.

With limited oil reserves and a greater demand for electricity, the world has reached a new turning point in the selection of energies. The world energy structure has gradually shifted from relying on oil and coal as the primary energy sources to developing multiple energies. For the purpose of environmental protection, people have recently advocated green and renewable energies in addition to nuclear energy.

In the modern time of human history, particularly since the beginning of the 21st century, we have seen a major scale-up in the use of hydropower, wind, solar, biofuel, and geothermal as energy sources. Commonly known as renewable energies, they are the driver of much of the growth in green energy. The present-day energy structure is an evolving spectrum of seven different kinds of energies composed of hydropower, thermal (coal, oil and natural gas), nuclear, wind, solar, biofuel and others (geothermal, ocean energy and marsh gas).

History demonstrates that economic prosperity is highly correlated with energy consumption. The economic growth of a country is heavily dependent on the increasing use of electricity. Electricity generation accounts for about

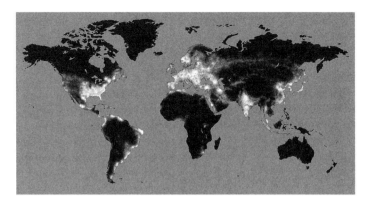

Figure 3 Global electricity utilization as indicated by nighttime electricity use, ORNL.

40% of the world's total primary energy consumption. However, among the world's 7 billion population, 2.8 billion have very limited access to electricity while 1.8 billion have no access to electricity at all. Those without access to electricity are not only deprived of the enjoyment of clean water and clean air, but also suffering from the worst health care, schools, and economics, and hence the shortest expected life expectancies. According to Oak Ridge National Laboratory, Figure 3 is the distribution of global electricity utilization in bright white as indicated by nighttime electricity use.

Energy Crisis

The world first started to take nuclear energy seriously after the energy crisis broke out in the 1970s, which led to a protracted global war in the scramble for oil resource. In the 1990s the burgeoning demand for oil throughout the world gave rise to a series of international disputes, and even to a war for oil. As a result, the highly reliable and low-priced nuclear power drew worldwide attention as an alternative. To provide a new energy that supports our life and economy is another contribution of nuclear energy.

However, nuclear power has been repeatedly challenged. The Three Mile Island nuclear accident in the US, the Chernobyl catastrophe in the former Soviet Union, and now the Fukushima disaster, raise many concerns. The social and psychological impact of these serious nuclear accidents is far more damaging than their actual environmental and human impact.

According to the book *Social Impacts of Nuclear and Radiation Accidents*, compared to the coal power chain (including coal-mining, transportation, the coal power plant and waste treatment), the nuclear power chain (including the nuclear power plant and its entire fuel cycle) is much less detrimental to workers and the environment (see Table 4).

Table 4 Table of comparative impacts of coal power chain and nuclear power chain.

Impacts		Coal power chain	Nuclear power chain	Coal/nuclear
To public health	Radiation	420 persons Sv(gWe-year)$^{-1}$	8.39 persons Sv(gWe-year)$^{-1}$	50
	Non-radiation damage	12 persons (gWe-year)$^{-1}$	0.67 persons (gWe-year)$^{-1}$	18
To workers	Radiation	90 persons Sv(gWe-year)$^{-1}$	8.91 persons Sv(gWe-year)$^{-1}$	10
	Pneumoconiosis	21.6 cases (gWe-year)$^{-1}$	4.4 cases (gWe-year)$^{-1}$	5
Loss of lives from severe accidents		35 persons (gWe-year)$^{-1}$	0.6 persons (gWe-year)$^{-1}$	60
To environment	Outflow	Obvious	Imperceptible/ Unnoticeable	
	Land taken up by solid wastes	2.1×10^4 m^2 (gWe-year)$^{-1}$	1×10^4 m^2 (gWe-year)$^{-1}$	2
	Land subsidence	1×10^6 m^2 (gWe-year)$^{-1}$	1.6×10^2 m^2 (gWe-year)$^{-1}$	6.3×10^3

Note: Quoted from *Social Impacts of Nuclear and Radiation Accidents,* 2011. UNSCEAR 1988

Table 5 Emission of greenhouse gas (GHG) by various energy chains [$gCO_2 (kW_eh)^{-1}$].

	Technology in the 1990s		New technology
	Maximum	Minimum	
Brown coal	1,233	958	837 (2005–2020)
Coal	1,310	969	756 (2005–2020)
Oil	903	804	547 (2005–2020)
Solar	280	92	27.6 (2010–2020)
Hydro	217	3.7	
Biofuel	56	28.3	
Wind	44	8.4	
Nuclear	19.2	8.4	

Note:
1. Quoted from *Social Impacts of Nuclear and Radiation Accidents*, 2011.
2. UNSCEAR, 1993.
3. Estimates by International Energy Agency (IEA) match the figures in this Table.

If we compare the greenhouse gas released in the life cycle of different energy chains (see Table 5), the nuclear power chain releases the smallest quantity of greenhouse gas.

Of the Spectrum of Energies, Which is Most Splendid?

Energy is no longer a luxury in today's world, but a necessity, just like rice or bread. We must strike a balance between energy supply, reliability and sustainability, and economic development (see Figure 4). Reliable energy supply should guarantee economic development and enhance people's quality of life.

In addition, reliability of supply is also key to energy. In the early morning of April 25, 2013, the Taiwan high-speed train signal system broke down. When the high-speed train operator decided to suspend the service for safety reasons, the affected passengers objected strenuously. Compared to this, it is not hard to imagine what kind of public outcry will result in the event of electricity breakdown in cities among the city dwellers who take electricity for granted. It underscores the importance of the reliability and sustainability of electricity.

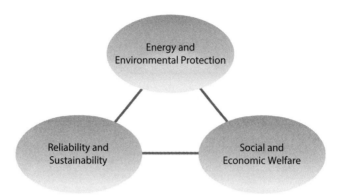

Figure 4 The necessity of energy and environmental protection.

Take the terrible six-decades-long record drought in the middle and lower reaches of the Yangtze River in China as another example. In April and May 2011, the lack of water for generating hydroelectric power resulted in 40% less electrical power generated in the central Hunan province in China compared with the same period in the previous year. But two months later in early June, the central region of China was ravaged by storms and floods. The capricious climate made it hard for the hydroelectric power plants to meet demands.

Unreliability of energy supply has its impact on sustainability. People prioritize hydroelectric power in order to protect the environment, but this energy source is limited and not sustainable. People then switch to heavy-polluting thermal power generators.

In the spring of 2013, the seasonal rain did not come and Taiwan suffered from an acute water shortage. Industrial and daily use of water was seriously affected, to say nothing of generating electricity with water. Even though a subtropical Taiwan is supposed to have plenty of rain water, we constantly face uneven seasonal water supply. Water shortages and an unsteady water supply are the norm in addition to such abnormal phenomena as badly silted-up dams and the frequent collapses of mountain slopes. Hydroelectric power is not a steady and reliable energy source for Taiwan.

Another case in point is ethanol, a biofuel produced from corn. Iowa, the largest corn-producing state in the US, started promoting ethanol as a substitute and additive for petroleum 20 years ago. But since large amounts of corn were used to produce the fuel, the price of corn for food and fodder rose sharply. These strategies, which are not favorable to people's livelihood and welfare, are far from viable solutions. Similarly, wind power, not sustainable in many places, generates low frequency noises which can often attract complaints.

Nuclear power plays a significant part in a spectrum of energies now, and this will continue for the foreseeable future. But unfortunately, there remain many obstacles to its development. Among the calls for improving the situation there are indeed some well-grounded concerns, but there are also many baseless social and psychological elements that are more akin to a child's irrational fear of the dark. People are haunted by the fear of nuclear disasters when in reality nuclear energy has a strong safety record. If the general public cannot freely and rationally explore the development of nuclear energy for human well-being and sustainable economic development, the issue will be confounded by political rhetoric and populist reactions, and therefore become much more complicated and unreliable.

There is an old joke about Ah Q and his wife, who keep two chickens at home. It so happens that one of the chickens is healthy while the other has become listless and sick. Ah Q's wife is not at all happy with this and hopes that her husband, who has had a rough time with his work and is between jobs, would do something to cure the sick chicken. When she comes home one night, she asks him what he has been doing. Pleased with himself, Ah Q tells her that he has bought some precious tonics and butchered the healthy chicken. Putting them together, he has cooked a nutritious soup which he is feeding to the sick chicken. He hopes that the sick chicken will soon become well and strong and his wife will be pleased.

Many people may think that Ah Q's act of killing the healthy chicken to cure the sick one is a joke. However, people like Ah Q can be found everywhere and the kind of things that he does are common occurrences around us.

Energy is a similar case in point. As the main electricity-generating fuel in the world, coal can be compared to the sick chicken. From whatever perspective, coal should be abandoned if we are at all concerned with the environment.

Unfortunately, our world has an abundance of Ah Qs. Instead of protesting against the pollution caused by burning coal to generate electricity, many people are eager to give up nuclear power probably for the purpose of saving thermo-power generation. Rarely do we see people give due attention to the severe industrial accidents brought about by the use of coal as an energy source. Insofar that they turn a blind eye to what goes on around them, they are no better than Ah Q.

Sense and Sensibility

Energy consumption is necessary for modern life. We should pay due attention to safety, environmental impact, and the reliability of energy supply for social stability and national security. We should also strive for an optimal balance between the competing demands of energy supply, social and economic well-being and reliability and sustainability when we consider energy and environmental protection.

To cope with the energy crisis, we must address the uncertainty of social psychological factors and the complexity of politics. We should adopt a rational and scientific approach to public policy-making. Such an approach will enable us to understand the tripartite relationship connecting energy, reliability and economic welfare in our society and appreciate the importance of reliable energy for sustainable economic development and improvements in quality of life.

To sum up, let me attempt a re-write of the poem "Reflections upon reading books" by Zhu Xi:

> *A spectrum of energies unfold in a fabulous case,*
> *Whereupon psychology and technology waver in two minds.*
> *Wondering which way to turn for the good of humankind,*
> *Reliable source lies in our gracious care flowing like a spring.*

As far as energy decisions are concerned, which is more important: sense or sensibility? Which energy is more glamorous? Perhaps neither wind (*feng*) nor solar (*guang*) are that "*fengguang*" (glamorous) as they might appear to be!

9

Facts about Background Radiation

The general public was thrown into panic in the days following the radioactive leak at the Fukushima nuclear plant. However, a lot of people are unaware of the levels of radiation they are exposed to on a daily basis. In fact, background radiation from the natural environment is much higher in Hong Kong than in other cities in Greater China or around the world.

Radiation can be roughly divided into two categories: non-ionizing radiation and ionizing radiation. Non-ionizing radiation refers to such low-energy radiation as light or radio wave which usually does not change the chemical property of a substance. What people are worried about is ionizing radiation which is often referred to as the background radiation. This kind of radiation has enough energy to dislodge the electrons of an atom and put them into a free state, thus producing charged ions, a process that is likely to bring about chemical changes to biological tissues and therefore is harmful to living things.

The data showing the daily average dose rates of radiation in different places in Taiwan are available on the website of the Atomic Energy Council of the Executive Yuan of Taiwan (see: http://www.aec.gov.tw/www/service/other/images/20100422-1.jpg). Though this kind of radiation may be harmful to health, there is no need to worry too much. According to the findings of epidemiology studies, only exposure to high doses of radiation will damage the human body. Low doses of radiation (i.e., natural background radiation within the safety range) will not cause harm to the human body or, at most, will pose only a very low risk. Residents of Hong Kong who want to know more about Hong Kong's background radiation can log on the website of Hong Kong Observatory (http://www.hko.gov.hk/education/dbcp/radiation/chi/r1.htm). Similar websites can be found in the US (http://energy.cr.usgs.gov/radon/DDS-9.html), Japan, France, mainland China and other countries and regions that provide information on the doses of current local background radiation.

Where is the Nuclear-free Homestead?

The discovery of trace plutonium in the soil near the Fukushima Daiichi Nuclear Power Plant on March 28, 2011 triggered alarm, but people's fears subsided the next day, presumably because the amount of plutonium found was within safety limits.

The natural background radiation in Hong Kong is many times higher than that of the average city around the world, even higher than that in Sendai, about 95 km from the Fukushima nuclear plant, after the nuclear accident. This is due to the cramped conditions in which most people live in Hong Kong, the dense residential areas, narrow streets, poor ventilation, the types of building materials and other sources. This is shown in the following table (Table 6):

Table 6 Background radiation in different places.

Location	Background radiation
An average city	0.05 micro Sv/hr
Tokyo, after the Fukushima nuclear accident	1.00 micro Sv/hr
Flight across the Pacific Ocean	25.0 micro Sv per time
Chest X-ray	10.0 micro Sv per time
Sendai, 95km from Fukushima, on April 21, 2011	0.20 micro Sv/hr
Inside a Hong Kong building	0.25 micro Sv/hr
Daya Bay nuclear power plant	0.05 micro Sv/hr
Taiwan No. 1, 2, 3, 4 nuclear power plants	0.06–0.07 micro Sv/hr
Taipei	0.06 micro Sv/hr
Alishan National Scenic Area, Taiwan	0.10 micro Sv/hr

The environment that we live in is full of radioactive substances which may emit ionizing radiation. Radon is the major source of background radiation in Hong Kong. Lurking in the air and being radioactive, radon can cause many kinds of lethal diseases. It ranks as the second most malevolent cause of lung cancer, next only to smoking (radioactive elements such as Radium (Ra) and Polonium (Po) are found in tobacco).

According to data issued by the IAEA, radon in cities mainly comes from buildings and decorative stones such as marble and granite as well as poor quality cement, ceramic and other building materials. Background radiation also comes from mountains and cosmic rays at the poles, or from underground radon, among other places. The water in a hot spring contains steam coming from the underground magmatic layer. As a result, it carries in varying amounts radioactive substances such as radon, and therefore has very small doses of radiation.

Hong Kong people like to tour Japan and I have often been asked whether a Japan tour is safe. Table 6 shows that as far as radiation is concerned, Japan is much safer than Hong Kong, unless you bathe yourself in a hot spring day and night. I wonder if many people know about this.

Cinders also contain radioactive elements such as thorium and uranium, and low doses of radioactive elements such as polonium, carbon, and so forth, find their way into wild mushrooms, leafy vegetables and even shellfish and other seafood popular with diners in this region.

Of course, a large dose of nuclear radiation can pose health problems but it is highly unlikely that someone exposed to such radiation will develop thyroid cancer, for instance, if he or she is outside the 30 km radius of a nuclear-stricken area or has not stayed in the area for hours on end.

Radioactive materials are not infectious, and food containing low doses of radiation will not affect one's health. In fact, radiation technology is already widely used for food sterilization and preservation, pesticides and medical treatment. The milk we drink and the vegetables we eat are all treated with radiation. The use of radioactive rays in medicine is even more extensive. Sometimes, a patient's blood is drawn and re-transfused to his/her body after the blood is treated with γ rays to destroy T lymphocyte activity.

Radiation can also help date ancient objects. The earth contains very small doses of radioactive substances such as uranium, thorium and potassium. The half-life period of these radioisotopes can be as long as hundreds of thousands years. The carbon-14 dating and the thermoluminescence dating based on radiation technology are widely used in geology, anthropology and archaeology.

According to the book *Social Impacts of Nuclear and Radiation Accidents* (by Pan Ziqiang *et al.*) published by the Media Company Limited of China Atomic Energy Press, "Even in the peak period, the equivalent density of radioactive substances such as I and Cs released during the accident of the Fukushima Daiichi Nuclear Power Plant is much lower than the density of Po and Pb in the air in China that comes from the burning of coal." Before we get worried about the impact of radiation in Japan upon mainland China, Hong Kong and Taiwan, we had better take a serious look at the fairly high doses of background radiation in our own environment.

Despite the various versions, the fictional and idyllic recluse called Shangri-la, well known in both Chinese and foreign legends, shares a common feature about its location: it is to be found high up in the mountains. Withdrawal into reclusion has been highly valued since antiquity by those who were bored with wordly affairs or wanted to escape the demands of public life. Therefore, the Peach Colony depicted by the great poet Tao Qian (or Tao Yuanming, 376 to 427 CE) during the time of the Eastern Jin dynasty (317 to 420 CE), is an ethereal utopia. Here comes an excerpt from his famous essay, *Peach Blossom Spring*:

During the reign-period T'ai yuan (326 to 397 CE) of the Chin dynasty there lived in Wu-ling a certain fisherman. One day, as he followed the course of a stream, he became unconscious of the distance he had travelled. … The fisherman, marveling, passed on to discover where the grove would end. It ended at a spring; and then there came a hill. In the side of the hill was a small opening which seemed to promise a gleam of light. The fisherman left his boat and entered the opening. (quoted from the following site: http://afe.easia .columbia.edu/ps/china/taoqian_peachblossom.pdf)

From these words, we can see that this Promised Land is also located in the mountains, and may even be in a big cave.

Judging from the information in Table 6, we may come to a bold conclusion — the background radiation in the idyllic worlds of Shangri-la or Peach-Blossom Spring is possibly far higher than that found in the cities. To illustrate this point, let's cite a poem written by Li Bai, one of the greatest poets of the Tang Dynasty, in praise of the beautiful concubine of the Tang emperor, Lady Yang:

> *Her face is seen in flower and her dress in cloud,*
> *A beauty by the rails caressed by vernal breeze,*
> *If not a fairy queen from Jade-Green Mountains proud*
> *She's Goddess of the Moon in Crystal Hall one sees.*

(Translation by Xu Yuanchong)

So you have to cheerfully accept the high doses of radiation whether you want to see the fairy lady in the Jade-Green Mountains or the Crystal Hall.

All this shows that we live in a radioactive environment all the time, including the nuclear fusion reactor, the Sun. Where, then, is our "nuclear-free homestead"? How can it be our ultimate goal to make our living environment "nuclear-free"? The advocates of "a nuclear-free homestead" are probably speaking out of ignorance, if they are not speaking in hyperbole.

10

Human Negligence of Occupational Safety Leads to Numerous Accidents

After the Fukushima nuclear accident, I pointed out that the railway authorities in China should pay attention to railway safety because the high-speed train network was developing at an amazing pace there. Not long afterwards, a serious collision occurred where one high speed train crashed into the one ahead of it. The puzzles behind the accident and the heavy casualties that resulted drew international attention.

Occupational Safety in Serious Jeopardy Right by Your Side

In the wake of the Fukushima accident, voices of fear were heard in Hong Kong about the threat that the nuclear power plant in Shenzhen might pose to Hong Kong. To prevent a similar incident from affecting Hong Kong, people suggested stopping the development of nuclear power in mainland China and Taiwan. In the meantime, the Taiwanese media reported that nuclear power plants along the Fujian coasts in mainland China would compromise safety in Taiwan, and therefore should be opposed at all cost. I wonder why nobody seems to take note of the fact that 25% of the electricity that benefit the residents of Hong Kong comes from the Daya Bay Nuclear Power Plant.

In order to set people's minds at rest, I ask friends of mine who have plans to travel to other cities to take with them a portable Geiger counter to measure the background radiation level of the places they visit. Their findings reveal that the background radiation level of Hong Kong is three to five times higher than that of Taipei or Beijing. It is almost the same as that of Sendai soon after the Fukushima accident, and is several times higher than that of the site of Daya Bay Nuclear Power Plant.

Like the proverbial pot that calls the kettle black, people tend to find fault with others while remaining blind to the imminent danger surrounding themselves, which, in this present case, has nothing to do with nuclear radiation. For example, according to a report by Nikkan Gendai magazine on November 21, 2013, improper bathing practice leads to 14,000 deaths in Japan each year. And, as reported by the *New York Times* of May 2, 2013, citing the released data by the Centers for Disease Control and Prevention, 38,364 people died from suicide in 2010 in the US, surpassing for the first time the 33,687 deaths from motor vehicle crashes.

"Occupational safety" is short for "occupational safety and health of workers". However, the term has taken on a wider meaning beyond the workplace alone to cover both private and public space. Be it a building, equipment or raw material, solid matter, dust or gas, anything in the environment that can lead to an injury falls under the scope of occupational safety.

An unsafe working place, a factory fire or any public peril that inflicts direct or indirect injuries or casualties is of great concern from the perspective of occupational safety. The burning of coal and oil to generate electricity results in the release of CO_2 that pollutes the air and damages the ecology, causing serious harm. The number of nuclear power plant incidents and the damage that they render are far lower than those caused by pollution and man-made incidents of the non-nuclear kind.

Even if we discount the psychological factor, the potential threats and dangers of non-nuclear occupational safety are frightening, yet our society has paid little attention to them.

The Unheeded Serious Problems of Occupational Safety

The earthquake and the tsunami in Japan sounded a warning bell, urging the public to pay strict attention to the safety of nuclear power and reflect on the harm of pursuing economic growth without due consideration to the overall well-being of society. Apart from protecting the environment, we should look closely at the problem of occupational safety.

On June 7, 2013, 47 people were killed and 33 others injured after a bus burst into flames in Xiamen in China's southeast Fujian province. Many passengers were trapped in the bus and burnt to death. On October 13, 2013, over 100 pilgrims were injured and dozens killed in a stampede on a bridge over a river leading to a Hindu temple in Madhya Pradesh in central India. Both accidents are related to issues of industrial and public safety.

On March 29, 2013, as many as 83 people were buried alive in a massive landslide at a gold mine site in Lhasa. On April 4, a high-rise residential building in the Indian city of Mumbai collapsed, killing 74 people. The roof of a supermarket in Riga, the capital of Latvia, a small country situated along the Baltic Sea, collapsed on November 21, 2013, killing 54 people. The man-made disaster was described as "mass murder."

A ruptured pipeline owned by China's largest oil refiner, Sinopec, caught fire and exploded in Qingdao on November 22, 2013, killing 48 people and affecting the lives of tens of thousands of others because of air pollution by the black smog from the explosion, and the contamination of water in the sea from the oil leakage. On April 17, a massive explosion ripped through the fertilizer works in Waco, Texas, killing more than 20 people and injuring 150. These are some recent additional examples of disasters.

These painful experiences should shake us from our complacency because it is difficult to be optimistic about the occupational safety in other lines of work in many countries. Workplaces are beset with dangers, and explosions and conflagrations in chemical plants are not uncommon.

Take Taiwan as an example. People tend to focus on the short-term benefits that a project can bring rather than its safety and durability. The completion of such hardware as a high-speed railway can be accelerated, but the assurance of safety standards and their effective implementation are not part of the Taiwan culture.

In our society, quality control mechanisms are far from perfect, and the problem of occupational safety is a reality as much as a potential danger. A construction site without proper safety protection is the norm rather than an exception. The collapse of the scaffolding of Taichung's Splendor Hotel in July 2011 is a case in point. According to statistics from the Taiwan Bureau of Labor Insurance, occupational casualties in Taiwan (excluding traffic accidents caused by drunken driving and driving without a license) averaged 30,000 to 40,000 every year over the last 10 years.

Train Collisions Throughout the World

Many disasters are the result of human errors. Among them, car accidents killing countless people all over the world are perhaps the most noticeable, leading us to believe that other kinds of traffic accidents are less frequent or serious because of better design, controls and management. We may even think the trains running on the designated tracks controlled by the modern computer and communication systems must be safe!

In fact, negligence in such areas as the well-established railways, subways, high-speed railway systems, construction and air traffic can also result in serious accidents, such as train collisions, collapses of buildings, and plane crashes. Disasters of this kind happen almost every day and are too numerous to count. Once an accident occurs, society rises in condemnation of human errors that have precipitated the disastrous events. However, as time passes, people become forgetful, and things return to their old routines.

In the name of speed and economic development, politicians take delight in bragging about the unprecedented scale of the projects they are undertaking and the record time that they set in completing them ahead of schedule.

Not long ago, China's high-speed railway was proud of setting a new world record of 486 km per hour. The high-speed railway network was predicted to reach 90% of the mainland's population and drastically accelerate urbanization. However, nobody was there to offer any guarantee of the safety and quality of its operation. In July 2011, the serious collision of two "Harmony" trains in Wenzhou, Zhejiang, killed 40 people, leaving us with a tragic example of the consequence of human negligence.

Six people have died and over 180 were injured when a packed French intercity train crashed into a station. The train from Paris to Limoges was carrying 385 passengers when it derailed at 5.15 pm on Friday, July 12, 2013 at Brétigny-sur-Orge in Essonne, 20 km south of Paris.

At least 79 people were killed and more than 140 injured in the passenger train in Spain on Wednesday, July 24, 2013, after all eight carriages of the Madrid to Ferrol train came off the tracks near Santiago de Compostela. It is reported that the Spanish train had been traveling at more than twice the speed limit around a curve. In the recent history of Spain, this is one of the worst rail crashes, among which a train derailed on July 3, 2006 on the underground system in Valencia killing 43 people.

The accident of July 24, 2013 also makes it the second major train accident in Europe since 2000. On November 11, 2000, a severe fire claimed the lives of 155 people in an Austrian funicular train because the electrical heater in the unattended cabin at the lower part of the train caught fire after the passenger train ascended into a tunnel.

The Washington Metro train collision occurred between two southbound red line subway trains during the afternoon rush hour of June 22, 2009. The train operator and eight passengers were killed and more than 80 were injured, making it the deadliest crash in the history of the Washington Metro.

Similar accidents also happened in Canada. A bus and a train collided in Ottawa on the morning of September 19, 2013, killing six people and seriously injuring eight others. In July, a Canadian freight train carrying crude oil derailed and exploded, killing 47 people in Quebec.

Train accidents in India are very common. Between 2007 and 2011, train accidents killed some 1,220 people, according to railway officials. A government-appointed safety panel in its 2012 report estimated that about 15,000 people were killed in India each year crossing train tracks in what officials describe to be an annual "massacre" owing to poor safety protection. As shown in a recent example, at least 37 Hindu pilgrims have died on August 19, 2013 after being hit by the Rajya Rani Express, while crossing the tracks at a remote station in Bihar.

The rail accidents and casualties so caused as listed in this section are only a very small sample of those occurred in the above countries and elsewhere in the world.

Natural Calamities are Less Destructive than Man-made Accidents

People like to brag about their superb skill and unusual courage in the face of perils, and it has become such a common problem in our society that public safety is greatly compromised. Almost without exception, the end of a typhoon in Hong Kong and Taiwan sees senseless casualties due to human negligence or faulty facilities or equipment. In a rainstorm or a typhoon, people are seen on camera stepping into the ocean in defiance of warnings and are engulfed by the waves as a result. A tourist coach forced its ways through Suhua Highway in Taiwan and ended up in a crash that killed many people. The consequence is no less tragic than the collision of high-speed trains in Wenzhou, China.

The macabre landslide that took place at the Seven Block Section of State Highway No. 3 in Taiwan in April 2010 buried three vehicles and killed four people. In April 2011, a fallen tree hit a forest train in the Alishan scenic area, overturning its carriage and killing five. The first accident was attributed to the fact that slope sliding monitors were not installed along the State Highway. As for the second accident, the authorities concerned failed to enforce the stipulation that overhanging trees and branches be pruned at regular intervals. Early in the morning of October 23, 2012, a fire broke out in a hospital in Tainan, Taiwan, injuring 75 and killing 12.

On New Year's Day in 1993 a tragic stampede in Hong Kong's Lan Kwai Fong entertainment area killed 21 people and injured six. The death toll was 41 with 80 injured in a fire at Garley Building in 1996. In 2003, 299 people died of SARS in Hong Kong.

Not long after the Fukushima nuclear accident, I showed some overseas guests around the waterfront of Tsim Sha Tsui in Hong Kong. As we watched ships weaving their way through heavy traffic in the harbor, I said to my friends that accidents would happen sooner or later if the government did not exercise strict control over the sea traffic. Just as I predicted, on the night of October 1, 2012, the most serious sea accident in the history of Hong Kong took place near Lamma Island. By October 6, 39 people had died and many others injured. The root cause of the disaster was more human-induced rather than natural.

According to the report issued by the China Economic Information Network, during the long eight-day holidays of the Mid-Autumn Festival and the National Day between September 30 and October 7, 2012, 68,422 traffic accidents took place in the mainland, killing 794 people.

Thus far, nobody has died of causes related to nuclear radiation from the Fukushima nuclear accident. Which kind of accident is more fatal, one cannot but ask?

In Taiwan, workplace safety is casually and carelessly maintained while its management lacks any accident-prevention mechanisms. The Taoyuan Airport was in chaos on a number of occasions. Planes turned into the wrong runways that have caused hundreds of deaths. In our society people have the bad habit of only looking after their own interests. Since occupational safety and environmental protection are not regarded as personal interests, it is not surprising that they are ignored.

The problem of occupational safety is much worse in mainland China. According to the *Blue Book of Cities in China: Annual Report on Urban Development of China No.5* published by the Social Sciences Academic Press in Beijing on August 14, 2012, China's urbanization rate broke through the 50% mark for the first time. The permanent residents of cities and towns in 2011 numbered 690 million, surpassing the population of regular rural residents. The urbanization rate is 51.27%.

With the rapid increase of population density in the urban areas, choosing sites for dams, high-speed railways and other big construction projects and maintaining their safety require special attention. High-quality equipment, qualified specialists and optimization of management mechanisms are three prerequisites to reducing risks. The potential threats of occupational safety do not come from the nuclear power plants alone. Japan's nuclear accident should serve as a useful reference to guide us in our management of all potential risks to public safety.

In order to ensure occupational safety, we should pay meticulous attention to the durability, consistency and stability of systems and facilities. The principles and methods described above are applicable to the safety and health of all tangible, intangible, direct and indirect aspects of public affairs. Even if our culture favors gambling, our society and government must act according to the rules.

By no means should anyone be allowed to gamble with or shortchange occupational safety or reliability.

11

When Will Environmental Pollution End?

In recent years, the environment has changed drastically, bringing about large amounts of pollutants as by-products.

The Misleading and Mislabeling Food Products

In 2012, horsemeat was discovered to have been sold as beef in Europe. On May 31, 2013 the skeletons of over 1,000 dogs were found at an abandoned site in Thailand. The authorities suspected that some unscrupulous meat merchants had used dog meat and then disposed of the skeletons.

In 2013, rice noodles of popular brands in Taiwan were found to contain no rice at all in their ingredients. Some stores in Hong Kong sold beef meatballs that had no beef in them. Equally shocking, one finds that bottled distilled water sold in the mainland was merely water taken from a river, and mutton was not used in the so-called spicy mutton hotpot. Low-grade meat was sold under the fraudulent label of premium steak in Japan's leading department stores, while chestnuts from China, posing as French produce, were put on the racks for sale.

Taipei-based bakery chain Top Pot Bakery, which does businesses across the Strait, was found in mid-August 2013 to be using artificial flavorings and fermentation-enhancement bacteria while labeling its products as using all-natural ingredients supported by a bioscience company. A local celebrity, also a co-owner of Top Pot Bakery known for her anti-nuclear stance, featured as its spokesperson in its misleading advertisements. It triggered an uproar in society for the extraordinary high price it charged for exorbitant profits. A beautiful story of all natural breads turned out to be a total lie.

In October 2013, Tatung Changchi Foodstuff Factory Co. Ltd, an international brand in Taiwan, was exposed to have used cheaper types of oil, additives and other ingredients to adulterate more than 80 types of its cooking oil products, including products labeled as pure olive oil, in order to save costs and to make exorbitant profits. Tests also revealed that the company's peanut and chili oil contains no peanut or chili in their ingredients. Such kind of fraudulent products has been put on sale in the market for years. The company is condemned for its blatant violation of food safety standards and total lack of concern for people's health.

In fact, cooking oil has always been a problem. Take the 1979 Polychlorinated biphenylsIn (PCB) poisoning event in Taiwan as an example. The rice bran oil was contaminated by PCB in the extraction process at a cooking oil plant in Changhua, Taiwan. As a result, more than 2,000 people were poisoned, causing multiple deaths and skin diseases. A further concern is that the toxic PCB cannot be released from the human body and may affect the next generation. Another prevalent phenomenon well known to people on both sides of the Taiwan Strait is the extraction of oil from used cooking oil which is then sold back to restaurants or roadside food stalls by illegal vendors for huge profits. The most scandalous event is the extraction of cooking oil from animal feeds which happened in 1985 in Taiwan.

Similarly, it was a common occurrence to discover that the coffee creamers in mainland China, Taiwan, Hong Kong, and elsewhere contained no dairy product. The process of fermentation of "smelly tofu" was, in fact, hastened by putting tofu in dirty water. Dried tofu was neither made of soya beans nor with fresh ingredients, but was still served to customers after being fried or cooked. There are too many similar examples to enumerate.

All these are enough to turn one's stomach and send shivers down one's spine when one first hears about them, but as long as they do not cause too much problem to our health, people don't seem to be bothered by them.

The Ubiquity of Biological and Chemical Sources of Pollution

The British Medicines and Healthcare Regulatory Agency (MHRA) issued herbal safety warnings and alerts on August 19, 2013 on its website, pointing out that the unlicensed herbal products like Niu-Huang Chieh-tu-pien and Bak Foong Pills contained high levels of lead, mercury and arsenic.

A report by the United States Renal Data System (USRDS), on the other hand, says that Taiwan has had for many years a high incidence of end-stage renal disease (ESRD), with a high percentage of people receiving haemodialysis and peritoneal dialysis treatments. It also ranks among the highest globally for many years in the prevalence of new cases, which is a result from ignoring food safety among others.

In early 2013, a surge in demand for formula milk powder among mainlanders caught Hong Kong by surprise. Before the government introduced curbing measures, local residents in Hong Kong experienced a shortage. What caused mainlanders to demand vast amounts of formula milk powder could only have been their lack of trust in dairy products manufactured in mainland China.

In 2010, test results showed that milk powder of the brand "Sanlu" contained melamine which might lead to the development of kidney stones in babies. The abusive use of plasticizer in beverages in Taiwan came to light in May 2011. Its damage to health is no less than, and even exceeds, the Sanlu milk powder incident. Over 500 daily food products were found to be contaminated by plasticizer, excessive intake of which would lead to hormonal pathological changes. In October 2012, Taiwan's Consumers Foundation tested samples of chrysanthemums sold in Taipei city and New Taipei city, and to the surprise of many people, it was discovered that all the samples contained pesticides. Prolonged intake of the contaminated floral tea is harmful.

Also in May 2011, an epidemic of enterohemorrhagic (EHEC) broke out in many European countries, causing numerous casualties and thousands of infections. Often, food pollution leads to severe social problems. According to the report released by the Centers for Disease Control and Prevention (CDC) in Atlanta, US, on June 7, 2011, over 120,000 people are hospitalized for food poisoning every year, about 3,000 of whom die.

Shanghai's waterways were clogged with dead pigs in early March 2013. Officially there were as many 11,000 carcasses. It is suspected that some of them might have ended up in local restaurants. By the end of the month, several people in Shanghai had contracted H7N9 avian flu, a virus previously unknown in humans. One of those who contracted the virus was a pork-trader. In the meantime, the bird flu epidemic continued to escalate in April in mainland China with more confirmed cases, causing worldwide concern. This public health problem could prove to be China's severest challenge since the SARS epidemic in 2003.

You are being too optimistic if you think such scenes of pig carcasses floating on the river are only seen in Shanghai. At about the same period, on April 5 to be exact, hundreds of dead chickens were found on an open ground near a pigsty in Chang Hua, Taiwan, with unknown causes. The way the dead chickens were disposed of is intolerable.

A report published in the February 2012 issue of China's *Environmental Science Journal* indicated that samples collected from China's six major water systems were found to contain a high level of environmental estrogen. The public casually dispose of environmental hormones and medicines, polluting ground and underground water and causing huge damages. Experts from China's Ministry of Land and Resources also revealed that 90% of its groundwater is polluted and the water in over half of its river systems is undrinkable due to its serious pollution. In May 2012, the incident of toxic capsules broke out extensively in China and affected nearly everybody. The world was shocked by the incident.

In addition, toxic substances have been in the food supply chain, for example pesticides, antibiotics, preservatives, and lean meat powder used in fish and agricultural products and processed food, all of which can often be lethal to humans. Recently, there are even reports to the effect that many newly produced garments and tools contain lethal toxics. Wearing these clothes could put people's lives at risk. In the Chinese communities across the Strait of Taiwan, there are countless cases of potential environmental pollution, a shocking and worrisome scenario.

Non-nuclear Pollution Permeates the World

In late May 2011, WHO announced that dry-cleaning substances and exhaust fumes from car engines are categorized as "carcinogenic particles".

As early as 1984, the leaking of methyl isocyanate from Union Carbide Corporation's factory in Bhopal of India, the largest-scale workplace accident in history, caused nearly 10,000 deaths and 500,000 injuries.

There is a recurrence of this kind of incident in history. Five major pollution incidents have taken place since the 1960s: the Amoco Cadiz oil tanker incident, the Gulf of Mexico blowout, the Cubatao's "Death Valley" incident, the India's Bhopal's incident and the River Rhine pollution incident. According to figures from China's State Environmental Protection Administration, in 2011, environmental damage in China, such as unsatisfactory harvests, forest recession and illness caused by air pollution resulted in a total loss amounting to 6% of GDP, i.e. RMB2,600 billion or approximately $410 billion.

Recently, the American Lung Association published a report that states "Particulate air pollution from coal power plants is estimated to kill approximately 13,000 people each year."

In January 2001, the Greek MV Amorgos grounded off the coast of Kenting, Taiwan, and caused a serious oil leakage, polluting 3.5 km of the coastal ecological protection area. In 2010, the Deepwater Horizon oil spill poured 5 million barrels of crude oil into the Gulf of Mexico.

In late March 2012, a serious natural gas leak occurred in the oil and gas field of the French energy giant Total located in the North Sea, off the northeast coast of Scotland, which took an estimated six months to stop. In the summer of the same year, Hong Kong was hit by "Vicente," a typhoon so strong that the observatory hoisted the strong wind signal 10 for the first time in 13 years. During the storm, seven containers on a cargo ship were blown out to sea. Six of them contained a kind of raw plastic material known as polypropylene. Around 150 tons of these plastic pellets were scattered on the sea surface and every beach in Hong Kong was polluted to some degree.

Also, according to a report published in Britain's Royal Society's *Biological Letters*, in the 40-year period from the 1970s to present-day, the size of the "Great Pacific Garbage Patch" formed by the convergence of plastic bags in the ocean has expanded over a 100-fold. Oceans full of micro-plastics seriously threaten aquatic life and oceanic ecology. The report shows that plastic is found in the stomach of one in every 10 fish in the Great Pacific Garbage Patch. As the oceanic garbage island covers a very extensive area, clearing it is difficult and costly.

On July 27, 2013, about 50,000 liters of oil spilled into the Gulf of Thailand from a leak in a pipeline operated by an oil and gas company, posing serious threats to the local flora and fauna. The navy warned that it could take a week to control and clean up the mess. The incident is the fourth major oil spill in Thailand's history.

Be it toxicity of plasticizer, the EHEC outbreak, poisonous gas leakage or plastic garbage, the examples are simply incalculable. Longitudinal studies in various areas in the past have revealed that damage caused by environmental disasters are extremely severe. Nevertheless, they do not attract the attention of the politicians. Neither do people march on the street in protest against them. The fundamental reason, I am afraid, is that no political advantage can be gained to show concern over these issues. It is better to turn a deaf ear to them.

Coal-fired Pollution Shocks and Global Warming

The traditional energies of oil, natural gas and coal constitute 87% of the world's overall energy consumption. More than 70% of electricity in China is generated with the thermal power that comes from burning coal and natural gas. In June 2011, China surpassed the US to become the country with the largest CO_2 emission. Yet, the electricity supply is uneven across China. While the eastern part of the country is suffering from a shortage of electricity, there is a surplus of it in the western part.

The electricity generated is not transmitted to the areas that need it, resulting in a situation similar to the frequent disruption of the electricity supply experienced by Taiwan and Hong Kong in early years. With disasters coming one after another, whence comes the resources to support their fast-developing industries?

In his book *The Carbon Crunch*, Dieter Helm at Oxford highlights the three core reasons for the global climatic problem, "coal, coal and again coal." At present, coal-fired power plants remain the economic pillar for many countries. According to the report by Britain's *Financial Times* published on November 21, 2012, the weekly increase in CO_2 emission in the two emerging countries of China and India alone is almost the same as the total emission of three new coal-fired power plants.

In January 2013, China experienced seriously heavy smog for several days in succession, preventing many people from going out. According to a report issued by the College of Public Health at Peking University, more than 85,000 people died prematurely of smog-related diseases in 2012 in the four cities of Beijing, Shanghai, Guangzhou and Xi'an. Also, in a December 2013 *Lancet* article by Chen Zhu, former Minister of Health of China, and his colleagues, air pollution causes 350,000 to 500,000 premature deaths on the mainland China each year. The main polluters were industry, coal and vehicles. This is believed to be a conservative casualties estimate which provides further evidence that CO_2 reduction is essentially a necessity.

Research result published in the journal *Nature Geoscience* pointed out that the Arctic Ocean is probably an important source of the powerful greenhouse gas "methane", the "cardinal criminal" in speeding up global warming. If that indeed is the case, the Arctic Ocean is going to increase by several folds the "positive feedback" of the global climatic system that has been confirmed to increase drastically the greenhouse effect, thus resulting in a vicious cycle.

Smog caused by fossil fuels is also found in Tokyo, Japan. In addition, on October 12, 2013, the temperature in central Tokyo reached 31.3 °C, the highest recorded temperature for October since records started being kept 138 years ago. It was also quite unusual to see Yoshino cherry blossom at this time of year. Normally, cherry trees begin to bloom in warm weather after a cold winter, and so it is difficult to imagine that they can burst into life in a cool October, according to a chief in the Sendai Meteorological Observatory.

CO_2 from coal burning brings about global warming. On May 10, 2013, a CO_2 monitoring station in Hawaii first reported that the concentration of CO_2 was standing at over 400 parts per million. The last time a similar level of concentration of CO_2 in the Earth's atmosphere was recorded was some 3 to 5 million years ago.

From the Industrial Revolution in the mid-18th century until now, the Earth's temperature has risen by one °C. If the current trend continues, it will rise one more degree in less than 100 years. CO_2 also causes the continuous thawing of the ice sheets in Siberia and North America, releasing the methane that has been sealed under ice for several million years.

According to an Assessment Report compiled in October 2013 in Stockholm, Sweden, by UN's Intergovernmental Panel on Climate Change (IPCC) on the basis of 9,000 scientific studies by 800 scientists, human activities were the dominant cause for global warming.

The report said global temperatures were likely to rise by at least 0.3 °C by the end of 21st century. However, the upper limit for the Earth and human beings is only 2 °C. It also predicted the world sea levels could rise by between 26 to 82 cm in the same period, posting a threat to coastal cities like Shanghai and San Francisco.

The Summary said it was "extremely likely" that human activities (e.g., use of fossil fuels) were the main culprit for global warming since the mid-20th century. That was an increase from "very likely" in the last report in 2007 and "likely" in 2001.

As global warming intensifies, even more methane will be released. Although the concentration level of methane in the air is very low, it has the capacity to capture 20 more times the heat from the sun than CO_2. The unusual methane emission that comes from the thawing of the ice sheets is indeed a cause for concern.

Excessive use of the highly polluting coal fire and severe energy wastage can cause global warming.

Global warming has led to the melting of ice layers at the poles and glaciers in other places, the rise of sea level and the flooding of low-lying areas. Extreme climates, such as rainstorm, snowstorm, drought and high temperature, have become frequent occurrences, causing enormous losses of life and property, crop failures, changes in the plant growth cycle, the extinction of different insect species and the frequent appearance of new epidemic diseases. Super rare EF5 tornadoes hit the Midwest of the US in May 2013, which merits our attention, too.

Environmental pollution has reached a point where we must face it squarely. Its damage to nature cannot be overlooked. In Taiwan, there were reports in recent years that nearly 65 km of the coastline along Kaohsiung are disappearing gradually, and, in the last 10 years, the Cijin coast has lost an area measuring 50 to nearly 100 meters wide and 3.6 km long, equivalent to the size of 30 football fields. The westward migration of the desert areas on the mainland of China is also a byproduct of ecological damage. Similar examples are too numerous to list individually.

Meanwhile, due to deforestation, the Amazon Basin, traditionally considered a bulwark against global warming may be becoming a net contributor of CO2, rapidly transforming and enhancing greenhouse effects. But not many people are concerned about that at all either.

Was the smog plaguing Beijing in early 2013 not shocking enough?

We cannot afford to overlook the phenomenon of global warming. We must exert ourselves to reduce carbon emissions. Otherwise, unimaginable serious consequences may arise someday when it is too late to rectify our mistakes.

Coal Mining

Even if one excludes the effects that coal burning brings to the environment and global warming as well as the many thousands of casualties in the world every year in the production and use of coal energy, coal-mining itself causes tremendous casualties. For example, coal-mining causes directly 54,000 deaths in China in the past 10 years.

Take Taiwan as another example. Its coal-mining history dates back 125 years, starting from the first official Western-style coalmine located in Badouzi, Keelung in 1876 to the closing of the last coalmine in 2000. For transportation and quality reasons, coal was mostly produced in the northern mountainous regions, north of Miaoli.

At the peak of the industry, an estimated 400,000 people in Taiwan depended on coal-mining for a living, and accidents were relatively commonplace. The casualties from such accidents in Taiwan topped the world at one time. On December 1, 1971, for example, 42 people died in a mining accident in Chishing Mine in Keelung, while three mining accidents that took place in Tu-cheng, Ruifang and Sanxia claimed 289 lives between June and December, 1984.

During the period of 1956 and 1971, more than 100 coal miners died during coalmining just about each and every year in Taiwan alone.

During these bitter and bloody years, coal-mining families suffered from debilitating work-related diseases; local residents suffered from dire living conditions, the regular loss of life, and serious air pollution; and there was a great deal of damage to the geological and ecological environment. The long-lasting damage from coal-mining and the consumption of coal in Taiwan and elsewhere lends credibility to Kharecha and Hansen's 2013 report, but no one seems to care how it has come to this or why.

The Melancholic Beauty of Idaho

The State of Idaho of the US is a place of enchanting natural beauty. The melancholy beauty of its fall is especially breathtaking, comparable to the "earthly paradise" of Shangri-La.

Records show that nuclear power was first generated on December 20, 1951 by a nuclear reactor near Arco, Idaho.

In the summer of 1993, I visited the US National Engineering Laboratory in Idaho, when a heated debate was in progress about the possible building of a local nuclear reactor. At that time the residents of the city of Idaho Falls were against the building of a heavy-duty nuclear-power facility. After conducting a detailed enquiry, I came to understand that the local people were not necessarily against the building of nuclear facilities that are bigger than the traditional nuclear power plant. Rather, they were opposed to what might happen after the reactor was built. The plant would employ thousands of workers, who would arrive with their families, together with the catering and other small-scale service industries that would add drastically to the population.

The increased population might give rise to more crimes, man-made pollution, organic or inorganic, traffic problems, social unrest and other issues, which would throw this "earthly paradise" into upheaval and destroy its serenity.

The residents of Idaho Falls did not necessarily mind an additional nuclear reactor with a limited number of staff and minor pollution. On the contrary, they were much more worried about man-made pollution.

Which Way Forward: "Nuclear" or "Non-nuclear"?

According to a public announcement recently issued by the US Environmental Protection Agency, the amount of mercury and suspended particles emitted from non-nuclear power plants is astounding. Its damage to human health goes beyond our commonplace understanding.

Coal cinder and soot not only produce acid rain and pollute the air, but also contain a large amount of radioactive thorium, uranium and other elements. In the year 2000 alone, the burning of coal worldwide released about 5,000 tons of uranium and over 10,000 tons of thorium. According to another estimate, the radioactive elements released into the air from coal burning in the US in 1982 are more than those from the 1979 Three Mile Island nuclear accident. Purely from the perspective of radioactive pollution, electricity generation by coal burning is extremely dangerous (see Table 4 in Chapter 8).

Non-traditional fossil fuels such as oil sands in Canada, pre-salt deposits in Brazil and shale oil/tight oil in the US have been discovered in abundance since the beginning of this century. Thus, in the next 10 to 20 years, the North American region may come to enjoy some relief from the pressure of relying on crude oil. It may even benefit from a stable and low-cost energy and become a net energy export area. Yet, the development of this new generation of fossil fuels will do nothing to reduce water and air pollution but in fact creates more pollution than traditional oil.

According to the cover story "The Truth about Oil" of Time dated April 9, 2012, the lifecycle carbon emissions of a barrel of oil sands that goes through the process of mining, refining and usage exceeds that of traditional crude oil by 10% to 15%. It can be foreseen that oil in the 21st century will be costly and will create more pollution.

In terms of reducing the death toll or controlling greenhouse gas emissions, nuclear power and renewable energy are both ideal, and nuclear power has made greater contributions to humankind than all the other energies in use, according to a report published in the March issue of the *Environmental Science & Technology* in 2013 by scientists Kharecha and Hansen from the US National Aeronautics and Space Administration (NASA). Compared with other similar reports, this academic thesis adopts a quantitative method for the first time to analyze the impact that nuclear power has had on human life.

They pointed out that, even taking into account the serious consequences of the three biggest nuclear disasters in history (discussed in Chapter 1), the benefit from the use of nuclear power between 1971 and 2009 have helped to prevent approximately 1.8 million deaths resulting from causes related to the use of fossil energies, especially coal. Kharecha and Hansen concluded that the losses brought about by non-nuclear energy use will be several hundred times greater than the threats posed to humankind by nuclear power.

On the basis of global projection data, Kharecha and Hansen predict that replacing fossil plants of burning coal and natural gas with nuclear power plants over the next 40 years could save 7 million more lives and 240 gigatonnes of CO_2-equivalent net greenhouse emissions. Therefore, they assessed that large-scale expansion of unconstrained natural gas use would not mitigate the climate problem but instead would cause far more deaths than the expansion of nuclear power plants. Their conclusion is consistent with Cohen's assessment on the impact of the environment on human life expectancy, as shown in Table 1.

What's more, during the above-mentioned period it was mostly the developed countries that had a high ratio of nuclear power use. India and mainland China had a very low ratio. In contrast, the major developing countries which relied heavily on fossil energies consumed an enormous amount of coal, natural gas and oil, which inflicted irreparable and immeasurable losses in terms of the environment and human life.

Emissions from new coal-burning power plants planned in Guangdong, China itself could cause as many as 16,000 deaths in the next 40 years, according to Andrew Gray, a private air quality consultant commissioned by Greenpeace. The long-lasting damage from coal-burning power plants in Guangdong again supports Kharecha and Hansen's report. Furthermore, smog in China becomes the most serious environmental problem in late 2013, to the extent that aircraft pilots are asked to perform instrument-guided landings. Some citizens in Beijing have to use GPS to direct them home and many today inhale more pollutants than those living in the Sahara desert.

All these remind us of the thick sulfuric acid mist of December 5, 1952 which brought London to a standstill for five days and is believed to have killed 4,000 to 12,000 people. Nowadays, air pollution from coal-burning may have become less severe there, but smog due to traffic fumes and others still results in more than 10,000 deaths each year in the UK. In fact, similar or worse smog due to "non-nuclear" pollution exists in just all of the world. To be sure, the nuclear accident caused by a major earthquake in Japan has brought to light the need to improve nuclear safety. But what is more important: to give up nuclear energy altogether or to cut down on the use of carbon? In the interest of environmental protection, should the priority be to eliminate "nuclear" or "non-nuclear" pollution? Environmental pollution and climate change have combined with chemical, biological, food-poisoning and man-made disasters to pose threats to our ecology. Why do we fail to come up with more thoughtful responses?

Why even up till now, our society still fails to come up with a comprehensive and objective environmental and energy policy?

Passengers and Taxi Drivers

Figure 3 indicates that the majority of the people in the world would be very satisfied as long as they could get frequent access to electricity. However, people living in these areas of scarcity with food, water and electricity have also to bear the adverse consequences of global warming brought about by our extensive use of fossil fuels, since 67% of the world's electricity is currently generated from thermal power, causing excessive carbon dioxide and sulfur dioxide emissions which lead to severe environmental problem. For those who enjoy cheap electricity and others who never have such a benefit, they will likely have to submit to the same disastrous consequences brought about by ubiquitous pollution caused by fossil fuel power generation.

For people who in their life time never have an opportunity to share our enjoyment of resources, and yet have to face a death caused by pollution created by the rich, do you call that fair?

Before we come to the end of this chapter, here is a true story as food for thoughts:

The taxis in New York are usually dimly-lit, old and dirty. One night, an alluring woman dressed in a super-short skirt boarded a taxi. The driver was a hefty black man, and he noticed from the rear-view mirror that his passenger, looking frightened, huddled against the door. He asked, "Are you scared?" The girl nodded her head. The driver said to her, "Take it easy. To tell you the truth, I'm afraid of you, too."

According to the crime record in the last 50 years, driving a taxi is a most dangerous job in New York City and taxi drivers face far more danger than passengers.

In his book, *The Travels of Lao Ts'an*, Liu E of late Qing dynasty writes, "In matters of this world, three to four out of 10 are messed up by wicked people whereas six to seven are by gentle people who are unschooled in the ways of the world." I thus have this observation: electricity used nowadays comes mainly from fossil fuel thermal power; and the number of casualties as well as the serious land and air pollution caused by the process of coal-mining, refinement and burning is shocking. Nevertheless, people in general do not seem to notice.

Mysterious is the way of the world, where truths and untruths are confounded. We find it difficult to tell who is more fearful in the story, the driver or the passenger. In the same way, before we lay the blame on nuclear power, we need to ask if nuclear power is really more fearful than other energies.

12

Non-nuclear Calamities
Are Also Horrible

The huge loss of life and property inflicted by countless natural disasters in history is incalculable. If we look at it closely, human factors are the major culprit of many serious disasters. While we may be able to protect ourselves from natural calamities, we are often unable to escape from the consequences of man-made disasters.

Natural Calamities

As recently as December 12, 2012, a powerful typhoon which hit the Philippines left 600 dead, 220,000 homeless and over 1,000 missing. Then on January 27, 2013, a fire at a nightclub in the city of Santa Maria, Rio Grande do Sul State, Brazil, claimed at least 232 lives. Even more devastating was the impact of Haiyan, the most powerful hurricane in recorded history of the Philippines, which hit the country on November 8, 2013, leaving 5,560 dead, 1,757 missing, and at least 14 million people affected including 1.8 million displaced children.

The dam failure of Banqiao Dam in Zhumadian, Henan province in China on the early morning of August 8, 1975, according to a Discovery

Channel report, caused more than 240,000 deaths due to flooding and the subsequent epidemics and famine.

Further back in the past, a wave of serious natural disasters – floods, droughts, storms and insects – struck China's Henan province, and led to a serious crop failure in 1941. In 1942, without a single drop of rain from spring to autumn, the provincial grain reserves were depleted. A major famine struck soon afterwards. By 1943, with pestilence spreading through the province, the conditions could not be worse. Within two years at least 2 million people were estimated to have starved to death. Meanwhile the Anti-Japanese War raged on. Man-made and natural disasters combined to add to the sufferings of the people.

Before the outbreak of war with Japan, mainland China was hit by a devastating flood in 1931. The disaster area covered more than 50% of the land and affecting some 50 million people, according to the data released by the Central Propaganda Department of the Kuomintang government. Although the statistics then was far from complete, by the most conservative estimates from local sources, no less than 3 million victims died as a result of floods, hunger and diseases. Thousands of kilometers of ravaged land with people dying of hunger left in the open is a pitiable sight not easy to bear. Incidents like these are not hard to find.

While or even if it is often impossible to avert a natural disaster, we can limit the damaging consequences by paying heed to human factors. This is clearly illustrated in the recent global economic turmoil.

Man-made Disasters

It is hard to understand why people paid so little attention to the looming financial crises around us, and underestimated the severity of their consequences and impact on people's lives and wealth, and on society as a whole, including its future.

In October 2001, the bankruptcy scandal experienced by Enron Corporation, the biggest energy trader in the world at the time, had an enormous impact on the US economy, financially ruining a great number of families.

The international financial crisis in 2007 and 2008 is regarded by economists as the most serious since the Great Depression in the 1930s in the US when over 13 million workers lost their jobs and countless numbers of people died of hunger. The international financial crisis in 2007 and 2008 brought about a global economic recession. The tangible and intangible losses were too massive to calculate. Only war is more devastating than a massive stock market crash. It is true to say that global finance got out of control towards the end of the 20th century, and many families around the world are still living with the dire consequences.

It is not uncommon to hear that a stock market crash was caused by human manipulation, bloodthirsty policy, insider dealing or collusion between officials and businessmen.

But more often than not, we are overly concerned about the threats of nuclear accidents, and fail to pay sufficient attention to prevent the colossal damages caused by natural disasters or to ensure the safety and reliability of our financial systems.

Horrible Consequences

Assuming that the most dreadful nuclear accident were to take place, as pessimists have predicted, would its consequences be 1/10, 1/100, 1/1000, 1/10,000 or even 1/100,000 as serious as the other tragedies in the 20th century, such as those caused by coal- mining accidents, bad habits (smoking or secondhand smoke) , natural disasters (the 1942 famine in Henan province), man-made calamities (the 1966–76 Cultural Revolution in mainland China), or wars (World War I & II)? Would the after-effects caused by nuclear accidents be more severe than the chemical and bio-logical pollution to which we are exposed every day? Obviously, there can be no definite answers to all these questions.

Are the various non-nuclear tragedies I mentioned above and in other places in this book not more worthy of our attention and studying to pre-vent them from happening again and again?

In this connection, I was reminded of a conversation between Confucius and his disciple Zilu in *The Analects*. "Zilu asked about serving the spirits. The Master said, 'While you are not able to serve human beings, how can you serve the spirits?' Zilu added, 'I venture to ask about death?' Confucius answered, 'What can you learn about death when you don't know every-thing about life?'"

This interesting conversation with its philosophical wisdom and logi-cal reasoning, exemplifies one of the tenets of Confucius' thinking and Confucianism after him, namely, one should not worship the gods and spirits blindly, and when there are the affairs of the human world to take care, one should not expend time, energy and money on other-worldly things.

The last line of Confucius' answer – "What can you learn about death when you don't know everything about life?" – suggests that when we have not yet figured out how best to live our lives, why do we bother to spend time on the business of life after death and what we should do with it?

If we extend this further, it might be said that we should not spend so much time and energy on matters that are illusory, far-fetched, and of uncommon occurrence. Things of this kind have a low probability of happening, and are serious only in our imagination. If we allow ourselves to be distracted by them, we will end up overlooking daily issues that are real and imminent and have a high probability of occurring, with disastrous consequences.

Even discounting the mortality brought about in the past by smallpox and the plague and nowadays by AIDS, the danger of nuclear power is miniscule compared to that of any one of the following: natural calamities, man-made disasters, infectious diseases, war, terrorist attacks, non-nuclear energy disasters, traffic accidents, industrial safety accidents, non-nuclear environmental pollution, financial crises, policy-motivated killing, suicide, food poisoning, the common flu and wrong verdicts.

In other words, while the demand for nuclear safety is reasonable, we should not neglect other man-made and natural disasters that have caused colossal damage to humans even as we raise tough questions about nuclear energy, especially when many of them can be avoided and their danger can be minimized. If we turn a blind eye to the accidents happening every day around us, and give up nuclear power with a complete disregard of its benefits to human well-being, we are, to paraphrase an old saying, foreswearing food for fear of hiccups or failing to see the forest for the trees.

In this way, how are we different from those people who "serve the spirits" before they know how to "serve human beings"?

Seeking Enlightenment Instead of Chasing Shadows of An Illusion

Do not seek to become a Buddha immediately in this life, but strive to become one in the next.

Many people do not like to applaud the strong who persist, but they are willing to shed tears for the weak who compromise.

Many people do not even know who their neighbors are, but they show great interest in and excessive concerns over whether there are extraterrestrial beings.

Many people feel satisfied with themselves for queuing up, but they routinely run red lights while driving and never yield for pedestrians.

Many people would not lift a finger to solve a problem at hand, but get all tied-up in a web of confusion over dubious occurrences and fabrications.

Many people speak openly and vacuously about environmental protection, but they wantonly despoil the environment, frantically seizing and wasting material and human resources.

It is said that three things can most excite the human mind: daydreaming, phobic fears and sexual fantasy. Wisdom comes from admitting honestly what one knows and does not know. If one knows little about the spectrum of energies, should one not, with due composure and humility, set out to find out more? Rather than matter-of-factly facing the challenge of nuclear safety, why should one consciously or unconsciously retreat to the world of daydreams, phobias and fantasies?

Many people do not seek to become a Buddha in this life, but they strive to become one in the next life. Why is it, I wonder?

13

Where Can We Find Safe Energy Sources?

Japan's unprecedented 9.0-magnitude earthquake and the tsunami that followed caused the shocking nuclear accident at Fukushima Daiichi Nuclear Power Plant. With the passage of time, we can now reflect on the enormous damage that a natural disaster can inflict, and the series of questions related to energy sources and people's deep concerns for global safety that the Fukushima accident has prompted.

Where to Look for the Spring of Energy?

In 2012, the US produced as much as 4,054 billion kilowatt-hours of electricity, of which 37% were generated by coal, 30% by natural gas, 19% by nuclear energy, 7% by hydroelectric power, 5% by such renewable energies as wind power and solar energy and the remaining 2% by oil and other energies.

Take California as an example. Southern California Edison announced on June 7, 2013 that it permanently shut down the troubled San Onofre nuclear power plant, capping an 18-month debate about the plant's future due to premature wear found on over 3,000 tubes in the steam generators.

Originally, the price of electric power was the same in Southern and Northern California. However, after the shutdown of two nuclear reactors in Southern California in early 2012, the local people have relied more on smart grids for electricity generated by natural gas. Consequently, Southern California power prices have persistently exceeded Northern California prices by 12%. The relevant data can be found in the US Energy Information Agency (http://www.eia.gov/todayinenergy/detail.cfm?id=10531).

The debate about if nuclear energy will replace other sustainable energies has become more heated. Partly because of the Fukushima incident, the German government announced on May 30, 2011 that all the nuclear reactors in Germany would be shut down by 2022 and a new energy policy will be drawn with the focus shifted toward the use of renewable substitute energies. This is considered as Germany's energy gamble.

In his 2012 State of the Union Address, US President Obama proposed in no uncertain terms that shale oil and gas in the US, of which there are rich deposits, should be developed to meet in part the nation's demand for energy resources.

As a matter of fact, the west of China is more abundant in shale oil and gas deposits. It is very likely China will replace Russia as the world's biggest natural gas exporting country in the future. The goal is set in China's 12th Five-year Plan to tap 100 billion cubic meters of shale oil and gas by 2020, which will take care of 10% of the total energy consumed in China.

In order to extract shale oil and gas, it is necessary to drill the rock stratum as deep as 4,000 to 6,000 meters underground. The released natural gas or oil can be extracted only after the fracking of the rock stratum with high-pressured water that is mixed with chemical substances. Such a drive for short-term solutions might cause bigger problems in the long run. This process of extracting shale oil and gas may shake the earth's foundation and bring about serious consequences.

In November 2011, a 5.6 earthquake occurred in Oklahoma, US, the magnitude of which had not been seen in 60 years. According to American experts, such an earthquake was closely related to the stratigraphic burst, a measure adopted in tapping shale oil and gas.

Two years after the 2011 Wenchuan earthquake in China, another serious 7.0 earthquake took place in the same province—Sichuan. Unfortunately, the exploration of shale oil and gas is suspected of causing these natural disasters.

New energy will give rise to new crises. The extracting of shale oil and gas will pollute water and the atmosphere and seriously damage the geological structure. The governments of France and Bulgaria have banned the extracting of shale oil and gas by adopting pressing techniques.

Most regions of the world are searching for new alternative energies in the face of the growing energy crisis sparked by fast economic development and the limited availability of the kinds of energy sources that human societies can use. Of these alternatives, nuclear energy has become one of the top choices for many countries. For example, the NRC approved in January 2012 the application submitted by the Southern Corporation in the state of Georgia for a license to build a nuclear power plant. This was the first approval that the NRC has granted for the building of a nuclear power plant since 1978.

In order to address public demand for environmentally-friendly energies, the UK granted on March 19, 2013 its approval for the building of a new type of nuclear power plant, to be constructed by Electricité de France (EDF) in the UK county of Somerset, paving the way for the building of a series of nuclear power plants. South Korea is also enthusiastic about the development of nuclear power.

Of the electric power consumed in Sweden, 49% comes from hydropower, 39% from nuclear energy, 8% from biofuel and only 4% from thermal power or others. On the whole, this energy mix can be considered as the cleanest and most environmentally friendly energy structure.

Germany is expected to abandon nuclear power by 2022. Thereafter, solar, wind and hydroelectric power will become the main sources of electric power. And thermal power will generate 20% of electricity.

Green energy depends on hydropower, wind, solar and biofuel. However, the development of biofuel is slow and has a long way to go. Hydropower is reasonably rich in the Nordic countries, but still insufficient to meet the industrial and domestic consumption demands. These countries have to rely on the installation of extensive solar panels, gigantic wind-turbines and complex transmission networks to provide supplementation and transmission. Such measures are increasingly encroaching on the fast shrinking green zones. In addition, capturing wind to generate power by wind turbines is causing depletion of the moist northern winds and weakening their capacity to reach to the south inland, thereby creating detrimental effects on the ecology which is contrary to the original objective of green energy.

The rising cost of green energy is pushing up the electricity bill, posing increasing financial burden to white collar workers and affecting the right of ten million people to basic livelihood.

Other countries began to reconsider their energy policies, too. France, which has an extremely low release of CO_2, was determined not to follow suit. The Minister of Environment of Sweden thought the decision made by the German government was "most unfortunate."

Part of the electric power consumed in Germany was supplied by the nuclear power plants of its neighboring countries, people cannot help feeling that Germany made its decision to abolish nuclear power at the expense of its neighbors. All these remind me of the scene captured in Wang Jia's poem *A Spring Day after Rain*:

> *Before the rain I still see blooming flowers;*
> *Only green leaves are left after the showers.*
> *Over the wall pass butterflies and bees;*
> *I wonder if springtime dwells in my neighbor's trees.*

(Translation by Xu Yuanchong)

Up till now nuclear energy is still the one with the highest cost efficiency from among a spectrum of available energies. The significance of this fact will become more and more obvious.

According to statistics issued after the investigation conducted by Germany in January 2012, Germany has to pay dearly for its decision to abolish nuclear power. In addition to having to rely on the smart grid for the supply of electricity from its neighboring countries, at least 20% of its enterprises have planned to move to other countries due to the inevitable great increase in energy costs after all the nuclear power plants are shut down.

According to a report in Reuters on January 17, 2012, Siemens estimated that Germany would spend at least $2 trillion more in its expenditure by the year of 2030 if all the nuclear power plants were shut down as planned.

Most other countries and regions just can't afford to follow the examples of Sweden and Germany because of differences in geography, resources distribution and economic conditions. Sweden is able to achieve an ideal energy mix, while Germany is pushing ahead with plans to abandon nuclear power. But neither cases can be readily emulated by other countries in the world. In other words, encouraging everyone to go vegetarian is not necessarily the best solution when at least half of the world's population (see Chapter 8) is still suffering from starvation and survival is of paramount importance. Likewise, for people living in poverty, even dim lights at night and a little food are sheer bliss.

Petroleum-producing Countries' Plans for Nuclear Power

The countries in the Gulf Cooperation Council, composed of petroleum-producing countries, have benefited from their petroleum wealth over the past 50 years. Today, with the rapid increase in their populations, there is a greater demand for electricity. Due to limited oil reserves and the soaring price of oil, these countries are now compelled to explore alternative energies. They concluded after a serious study that nuclear power was superior to alternative energies. As a result, since 2006 the Gulf Cooperation Council has been supporting the development of nuclear power.

Several other countries are looking at developing their nuclear power industries, too. In 2010 Kuwait and Venezuela were seeking the assistance of France and Russia respectively for vigorously developing nuclear power in their countries. Though the plans had been suspended in both countries after the Fukushima nuclear accident, the topic was brought up for discussion again in Venezuela in 2013. Meanwhile, the government of the United Arab Emirates has signed an agreement worth $40 billion with a Korean corporation to design, build and operate four nuclear power plants. Saudi Arabia is seeking the help of the Obama Administration of the US for the same purpose.

On February 19, 2013 Jordan held discussions with Russia in Moscow about preparatory work for the building of a nuclear power plant. Indonesia is planning to build its first nuclear power plant, located in Central Java, with official operation set for 2019. Assisted by Russia, Iran, a major petroleum-producing country, opened its first nuclear power plant on September 12, 2011, and is making preparations for the building of more nuclear power plants.

The US and Russia are known for producing petroleum and generating nuclear power. All the above-mentioned major petroleum producing countries, except Kuwait and Venezuela, are making energetic efforts to develop nuclear power. They can be regarded as newly emerging countries, as indicated in Table 7, using nuclear power in addition to the countries listed in Figure 1.

Table 7 Countries using and planning to use nuclear energy.

Current status	Countries
Using nuclear energy with plans to build more	Argentina, Bulgaria, Brazil, China: mainland*, Finland, US*, UK, India, Iran*, Korea, S, Pakistan, Slovakia, Russia*, Ukraine, Czech Rep
Using nuclear energy with no plans to build more	Armenia, Canada*, Hungary, Mexico*, Netherlands, Romania, Sweden, Spain, Slovenia, South Africa, France
Using nuclear energy with plans to review nuclear energy policy	Japan, Taiwan

(Continued)

Table 7 (*Cont.*)

Using nuclear energy with plans to phase out nuclear energy	Germany, Switzerland, Belgium
Not yet using nuclear energy with plans to build nuclear reactors	Bangladesh, Belarus, Chile, Egypt, Indonesia*, Israel, Jordan, Kazakhstan, Korea, N, Lithuania, Malaysia, Poland, Saudi Arabia*, Thailand, Vietnam, United Arab Emirates*, Turkey, Nigeria, Venezuela*

* Oil producing countries
Source: World Nuclear Association.
2013 updated information at: http://world-nuclear.org/Information-Library/

World-wide Trend in Building Nuclear Power Plants

According to Organization for Economic Co-operation and Development's Nuclear Energy Agency, a recent survey conducted by the intergovernmental organization of industrialized nations found that 26 of its 34 member nations planned to build more nuclear power plants.

Some 64 reactors were under construction before March 2011, after which time the number of units under construction had risen 10% globally despite the Fukushima accident. Because the modern nuclear plants have a designed operating life of 60 years, instead of 40 years of their previous generation, it makes the new plants even more cost-effective.

According to Fuel Cycle *Stewardship in a nuclear renaissance*, published by the Royal Society in October 2011, "Many countries have expressed an interest in nuclear power as a major component of their climate change policies and to address their energy security needs. This includes both countries with existing nuclear power programs, as well as countries embarking on programs for the first time.

"The construction of new reactors is likely to be limited in Europe and the US… Construction of new reactors is further advanced in South and East Asia, especially China, India; and South Korea, as well as Russia (see Figure 5). These countries are likely to lead a global expansion of nuclear power: the so-called 'nuclear renaissance'. The Middle East could emerge as the second largest market for new reactors. "

The same report projected that 300 to 800 completed nuclear reactors will be in operation in the world by 2029. Given the uncertainty of projections—the more deviation from the average, the more distant the

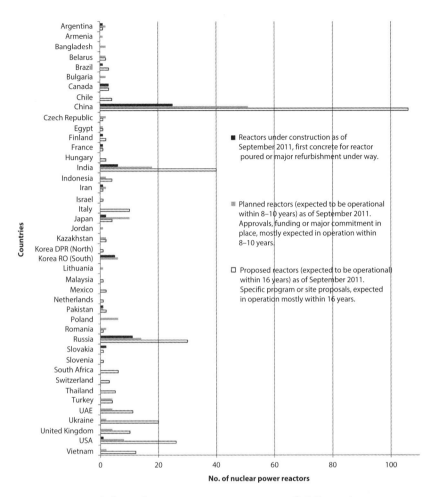

Figure 5 Proposals for nuclear power reactor construction (WNA 2011).

projection is—it is likely that 600 nuclear reactors will be running in the world in 15 years' time, taking into consideration all possible factors. Since we have 400 reactors operating after the Fukushima accident, 600 reactors would be considered a relative growth. See the projected scenarios in Figure 6.

As for the efficient use of energies, we have to understand that energy conservation is only the first step. We therefore have to go beyond energy conservation alone. In view of the general trend of developing nuclear power and the limited availability of nuclear fuel, in addition to transforming nuclear weapons to non-military use, we must optimize the exploration and tapping of a full spectrum of energies.

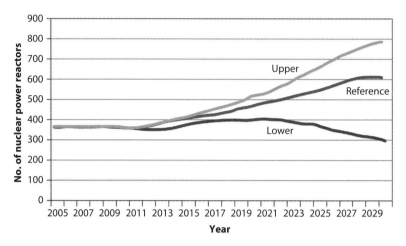

Figure 6 Worldwide nuclear generating capacity scenarios (WNA 2011).
Source: World Nuclear Association (2011) http://royalsociety.org/uploadedFiles/
Royal_Society_Content/policy/projects/nuclear-non-proliferation/
FuelCycleStewardshipNuclearRenaissance.pdf

Essential Conditions for Safety Design of a Power Plant

The tragic scenes after the nuclear accident in March 2011 remain fresh in our memory. We are obliged to think of the future development of energy sources in a rational way and reconstruct what happened to the Fukushima Daiichi Nuclear Power Plant.

As I mentioned earlier in the book, the Fukushima power plant is not the nuclear power plant nearest to the epicenter of the earthquake. In fact, the Onagawa Nuclear Power Plant is nearer than the Fukushima Daiichi Nuclear Power Plant. However, the Onagawa power plant operated normally and there was no radioactive leak there. Confronted with 15-meter-high waves, the three nuclear reactors of the Onagawa power plant remained intact simply because it was built on a hillside slope.

The safety of all energies, including nuclear energy, is determined by the following three essential factors: the hardware, the human factor and system management. In most cases management and the human factor are of prime importance. There is no exception in the world's three major nuclear accidents.

From the lessons that we draw from the Fukushima nuclear accident, we have reexamined and revised the specifications of nuclear safety, modified the report procedure and raised the design standards of nuclear reactors.

All over the world, strict measures are now taken to exercise the quality monitoring and control of a nuclear power plant.

The fact that the contingency generators of the Fukushima nuclear plant were inundated by water taught us a hard lesson: contingency generators must be installed at a higher place or in a watertight generator room. Besides, the fact that there was no other choice but to use seawater to cool the nuclear reactor prompted people to propose a series of measures to improve the cooling system. For example, back-up generating trucks should be deployed nearby. Improvement should also be made to ensure that in an emergency people can operate without electricity all the components of the cooling system.

Apart from nuclear power plants, the same strict safety standards should apply to the design of power generating devices using other energy sources. This ought to be another lesson and revelation that we learn from the Fukushima nuclear accident.

Ah Q's Story of Generating Electricity by Marsh Gas

The Fukushima nuclear accident drove people in the world to reexamine the future of nuclear safety and global safety. However, I have to stress that there is also a need to strike a balance among consumption, reliability, cost benefit and sustainability of other energy sources.

Even if we put aside the ecological impact of building dams over rivers, fickle weather makes it a challenge for people all over the world to run their hydropower stations efficiently. The biofuel ethanol that is made from corn is also beset with problems. Though ethanol can serve as a substitute and additive for oil, the use of corn in large quantities to produce fuel will give rise to a hike in its price, thus exacerbating the world shortage of grain. I don't think this is a solution that wins people's heart.

Solar energy and wind-generated energy have their own unreliable factors, because these two energy sources are affected by changeable weather. Coal used as fuel causes serious pollution detrimental to people's health and coal-mining inflicts heavy casualties and kills tens of thousands of miners in the world every year. Additionally, more than 70% of China's railway transport is devoted to the transportation of coal. Firstly, the transportation itself is energy-consuming and low in efficiency. Secondly, how ironic it is to have to use up so much of the already crowded transport network to transport coal!

This reminds me of another story about Ah Q and his wife. When they find out that marsh gas can be used to generate electricity, the couple are amazed at the unlimited magic of the marsh gas pit where excrement is left to ferment to produce electricity and heat. Being energy-saving, safe and environmentally friendly, marsh gas is a cheap energy source that is used widely to generate electricity.

Thereupon, Ah Q encourages his family members and visitors to eat more and drink more. In the meantime, he and his family breed a large number of domestic fowls and animals. Ah Q also lays down the general rule that waste should be utilized and excrement should be collected and processed. In order to produce more waste and excrement, Ah Q purchases a large amount of food and feed. He intends to produce the largest amount of bio-energy on schedule so as to generate electricity. A year goes by. The money that he spends on producing waste far exceeds the money that he saves in the power bill, but Ah Q is very pleased with the result of his energy-saving efforts.

"It takes time for nature to provide us with our limited resources and yet human desires are infinite. To satisfy such infinite desires with a limited supply in the absence of any effective principles and mechanisms to regulate the process, we will end up wasting our resources and become all the more impoverished for it."

These words from the Tang poet Bai Juyi should help us understand Ah Q's efforts in a new light.

Don't we see Ah Q and his many clones by our side who, in the process of transforming energy, only end up wasting more of our resources? Related to this, it is reported that some public toilets in North Korea are usually locked for fear people will steal the feces for manure. The Dutch are collecting urine donation for using as fertilizers, among many other countries in the past history.

The Need for Rationality

It is understandable for some people to fear nuclear power. Electricity has become a basic necessity in our daily life, and the demand is increasing. Any sensible person will agree with the following conclusion: if humankind cannot avoid the issue of nuclear power, their only way out is to seek an acceptable solution that can balance the benefits of nuclear power with the risk of nuclear radiation. One has to weigh carefully the gains and losses in making an overall plan.

Unfortunately, among the clamoring voices, we seldom hear cries of worry that are substantiated with valid reasons and factual arguments. Rather, most people hold the opinion that public responses and psychological reactions are always reasonable and correct. The public have to approach the question of energy strategy rationally, on the basis of comprehensive scientific knowledge and with a view of furthering human well-being and economic development. Otherwise, this debate will stray from the real issues, and be misled by ideological biases or sophistry.

To address the energy crisis, we must strengthen public participation and build up their trust in energy policies. Only in this way can we solve the related social problems. It is of vital importance that government control is effective, information transparent and the supervisory bodies competent if we are to reach the goal of effective energy management.

Both private and public investments are needed to generate creative ideas for solving energy problems. At present, no perfect solutions have been found to meet the challenge posed by the energy situation. Based on the principle of collective responsibility, we may designate a portion of the fees collected from electricity consumption for the improvement of safety measures and professional training so as to enhance reliability and raise standards.

The Fukushima nuclear accident has once again confirmed the well-known saying: "Crises come with opportunities." Recently, a spectrum of energy alternatives have been extensively explored (Table 5). In regard to nuclear energy, beyond the conventional uranium-fuel nuclear reactors, thorium nuclear reactor, which has been in existence for decades, is now given a new look by China, Norway, UK and others. Thorium fuel cycle, if successfully developed, could provide clean and economic energy with minimum nuclear waste although reliability of such a technology remains in question.

A reliable energy source can not only propel the sustainable development of the economy, but also improve people's living standard. We should consider energy development in a comprehensive way. Rather than using polluting and unreliable energies, we should reduce gradually our dependence on coal-generated power, and give priority to the production and use of clean and reliable energies. In the meantime, we should draw up and implement thoroughly a policy to encourage and reward efforts in conserving energy and increasing energy efficiency, because energy and environmental protection are closely related to social well-being and global safety.

From the broader perspective of energy security, the right direction is to reduce carbon emissions, clean up the environment and avoid energy choices which produce global warming.

14

Pick up Our Share of the Energy Cost

Hot topics like energy conservation and environmental protection often end up in empty talk and lip service. One example is that in charging for the use of water and electricity, no thoughts are given to environmental protection. That the charge does not provide any incentive towards conservation is another. The value of existing forms of energy comes with a price, that of threats to the environment and our safety. For the sake of the well-being of society, it is necessary to work out a comprehensive plan in our search for energy sources that are economical, efficient, safe, clean and environment-friendly.

Hike in European Electricity Prices Expected

Over the past four years, household and industrial consumption of electricity across Europe has increased by 17% and 21% respectively. In the meantime, British government's predictions of sharply increased electricity prices in the next decades are getting renewed attention. A government report found that with current policies subsidizing green power, British electricity costs will rise 33% by 2020 and 41% by 2030.

On October 10, 2013, the chief executives of 10 energy companies accounting for half of Europe's electricity production called for an end to subsidies for wind and solar power. They also called on the European Union to develop a system to reward companies that maintain standby reserves, to compensate for power shortfalls when the wind doesn't blow or the sun doesn't shine.

How Much Does Energy Cost?

To use energies, especially those for generating electricity, is to bear the actual tangible and intangible costs. Moreover, power-generating energies continue to create pollutants with potential dangers to the environment and the people who have to work with them. They come with related costs, which can be termed "expected costs", expressible in the life-cycle costs of the energy in question. Taking into consideration the concerns of environmental protection, the details of the overall cost of electricity are as below.

The operating principles are applicable to any of the known and unknown energies in the whole spectrum of energies:

1. The costs of research and development, investment, exploitation, generation and transformation of a particular kind of energy;
2. The costs of the construction of the power plant and its specific operating expenses;
3. The various investments and expenses for the power plant to deal with long- and short-term environmental pollution under normal operating conditions;
4. The costs of environmental protection of the power plant, i.e., the expenses on restoring the damage done to the environment by an accident, including the repair expense, and what is required to restore the ecosystem after the repair;
5. The tangible and intangible societal costs, actual or perceived, of a particular kind of energy. Some of these can be reduced by education, others will increase due to political, social and psychological disturbances caused by human factors;
6. The opportunity cost of overlooking better energy alternatives, resulting in lower social benefits or wrong policy decisions;

7. The various costs of preventive maintenance, overhaul and warranty for all hardware and software in the plant during its life-cycle;

8. The costs of making regular compensation to the neighboring communities and environments inside and outside the power plant and the sites of energy extraction. In addition, we should include the costs of manpower training, public education and community promotion;

9. The costs of maintaining national and social security through the effective management and efficient use of our energy resources. We need to also consider the opportunity cost for the use of the energy sources.

A comprehensive and rigorous evaluation shall be carried out for the above costs and investments with due regard to the odds and probability of accidents. Only after comparing the overall life-cycle costs of various energies can we make a choice as to which type of energy to use.

According to the estimate of Patrick Momal of Eurosafe, the radioactivity-related expenditure after serious nuclear accidents accounts for only 20% of the overall life-cycle costs, while the rest are incurred by other factors within the life-cycle. I thus thought of an old trite, about freedom by P. Sandor, which I have modified below to describe an energy's life-cycle cost:

> Life is dear but love even dearer,
> Yet, both can be cast away for the sake of freedom.
> Costs of energy can be calculated,
> Risks can be anticipated.
> And the life-cycle cost?
> Something we cannot afford to ignore.

The Real Cost of Electricity

Whether due to the real risks involved or psychological discomfort, many people want to be far away from a power plant even when they are enjoying the convenience it brings. With increasing distances, the chance of being affected by pollution and one's fear of it diminish. As there are pollution and other potential dangers associated with the plant and its surrounding facilities (e.g., dams and windmills), we can use the smart grid to regulate the supply and hence the price of electricity. In determining how much to charge the consumer, we can take into consideration the cost of the risk, in addition to the production cost, the amount of electricity and the time when the electricity is used.

With the help of the grid, the electricity that is used does not come from a single energy source but from a number of sources through various combinations of transmissions. The payment from the user has to reflect the potential risks associated with different power sources. For example, if a user's electricity consumption comprises 70% thermal power, 20% nuclear power and 10% hydropower, we should consider the percentage of three types of energy sources when calculating the cost of electricity. As for the heavy polluting thermal power, we can charge a relatively higher rate; for the renewable energy coming from green sources, we can allow a discount based on the costs of generating electricity by these means.

In addition to conserving electricity, we may also take part of the revenues derived from various energies to finance the research and development of new energies and environmental protection. Innovation, the fundamental key to sustainable development, has not been given sufficient attention by our traditional culture, but we can work towards it at a policy level. Given the different levels of environmental pollution caused by different energies, the percentage of research funding levied on thermal power should be relatively higher.

The ideal condition is for the consumers to declare what combination of the various sources of electricity they are willing to accept, and the power company will then charge the consumers on the basis of the declared percentage, the amount of the electricity consumed and the guidelines mentioned above.

The government may also work out strategies of energy developments on the basis of the declared percentage from specific units and individuals. Only by doing so can we see the extent to which the public attaches importance to environmental protection, the discussion of which could then rise above empty rhetoric to become one with real substance. We should use the overall choice of power sources as a reference for framing and implementing energy policies, and adjust and develop step by step. Sweden has considered implementing some of the measures listed above. But such arrangements depend on a national consensus on environmental protection and the government's determination to carry out relevant policies.

In terms of the pros and cons and risks of the spectrum of energies, the time is premature for a sensible discussion to take place where people have yet to come to grips with the crux of the problem, namely, the idea that nothing comes free. But in view of the current cultural and national conditions, in some regions which are plagued with energy deficiency, we can at least adopt a realistic method of rewarding the efficient users with a discount. Big users, whose consumptions exceed the limits on the other hand, will see their bills rising by a geometrical progression.

The cost we pay for using electricity must take into account the economy, the costs of electricity generation, transmission and transformation, the sustainable well-being of the environment and safety and related social, political and psychological factors. Even the gains from energy conservation have to be taken into consideration. All these add up to the true cost of electricity.

Prior to setting electricity prices, we must think about risks, both long-term and potential, social responsibilities and the availability of sources. After all, we have to pay the price for possible crises. The risks attached to energy use should be shared by everyone. Please see Table 8 for the pollutions caused by all sorts of energy during electricity generation.

Stop Food Waste

Let's use food disposal as another example of wastefulness. From 1999 to 2010, in addition to construction and industrial waste, Hong Kong people generated about 3,000 tons of food waste every day, which accounts for one third of the municipal solid waste. According to data extracted from the local Environmental Protection Department, three-quarters of the food waste comes from household food waste.

In 2012, the local household waste was about 6,000 tons per day, adding up to 45% of the total landfills. A fact sheet reports that the number of discarded moon-cakes reached 2.8 million in 2008 alone. Another report indicates that Hong Kong people produce 2.6 kilograms of waste a day per person, almost the highest in the world, even higher than countries like Denmark (2.2 kilograms per capita) and the US, which are the highest and 2nd highest member states in Organization for Economic Co-operation and Development (OECD), respectively. It is about one kilogram higher than South Korea and Taiwan.

Kitchen waste and food waste in the big cities of mainland China are even worse than that in Hong Kong. Reducing food waste can lead to less food consumption, and less human, material and a multitude of other energy sources that go to collecting, processing and cooking the food. It is really a pity to see so much food wasted in Hong Kong and mainland China where the wealth gap is equally serious. Mencius comments on the disparity of wealth in his days in this way: "Though there is fat meat in our kitchen and there are fat horses in our stables, the look of hunger is on our people's face and bodies of people who have died of famine are lying in the wilds."

Table 8 Emissions produced by 1 kilowatt-hour of electricity based on life-cycle analysis.

Generation option	Greenhouse gas emissions (in milligrams CO_2/kWh)	Sulfur dioxide emissions (in milligrams SO_2/kWh)	Nitrogen oxide emissions (in milligrams NO_2/kWh)	NMVOC (in milligrams / kWh**)	Particulate matter (in milligrams /kWh)
Hydropower	2–48	5–60	3–42	0	5
Nuclear	2–59	3–50	2–100	0	2
Wind	7–124	21–87	14–50	0	5–35
Solar photovoltaic	13–731	24–490	16–340	70	12–190
Biomass forestry waste combustion	15–101	12–140	701–1,950	0	217–320
Natural gas (combined cycle)	389–511	4–15,000*	13–1,500	72–164	1–10
Coal – modern plant	790–1,182	700–33,321	700–5,273	18–29	30–663

* The sulfur content of natural gas when it comes out of the ground can have a wide range of values. If the hydrogen sulfide content is more that 1%, then the gas is usually known as "sour gas." Normally, almost all of the sulfur is removed from the gas and sequestered as solid sulfur before the gas is used to generate electricity. Only in the exceptional case when the hydrogen sulfide is burned would the high values of sulfur dioxide emissions occur.

** NMVOC means non-methane volatile organic compounds.

"Hydropower-Internalized Costs and Externalized Benefits," F. H. Koch, International Energy Agency (IEA), Ottawa, Canada, 2000. www.nei.org/resourcesandstats/documentlibrary/protectingtheenvironment/graphicsandcharts/lifecycleemissions/

When we look at the extent of food waste in our society, can we stop similar thoughts from arising in our mind?

Sorting food waste in Taiwan is beginning to work. Not only do households begin to separate organic from inorganic waste, in some buffet and fast-food restaurants, the diners volunteer to separate food waste from other garbage when they return the trays. Some people even bring their own chopsticks to restaurants in order to use less disposable chopsticks.

However, for a great part, the world-wide food problem is also a distribution problem. Sorting and recycling the leftover food is not necessarily better than controlling the intake of the food, which is certainly not better than the appropriate distribution of the food. Stop food waste!

Insatiable Desire for Energy

Examples of waste of resources are everywhere. Take the air-conditioning in public places in Hong Kong for example, which can easily be turned up by five °C. Many office buildings are left brightly lit after office hours, even when no one is doing any work inside. In addition to being a waste of energy, the light pollution of commercial neon lights is harmful to the ecology. As long as we make an effort, we can easily reduce power consumption in Taiwan and Hong Kong by 10%.

At the time when we are using up the world's energy, our effort in cherishing and conserving energy sources has by contrast been disappointing. There are many times when people put on publicity shows. For example, the global environmental protection relay event "Earth Hour" reached Nanjing, mainland China on the evening of March 24, 2013. The data of Nanjing Electric showed that at 8:40 that night the grid load jumped unexpectedly from the highest record of 4.85 million kWh to an embarrassing 4.90 million kWh. Also in mainland China, the number of vehicles had already reached 200 million in 2011, far surpassing the estimated number of 100 million vehicles for 2020. It is now using 450 million tons of crude oil per year, crossing the red line of oil consumption 10 years in advance of what is estimated for 2020. Failure to control China's oil consumption will bring great pressure on the whole world in the future.

Amory Lovins, an American energy policy expert, states that the US economy could grow to 2.6 times its present size, get completely off oil, coal and nuclear energy and use one-third less natural gas by 2050, if it could reduce the waste of energy at homes, in offices, factories and vehicles and improve energy efficiency.

Perhaps because people are content with the status quo, they are unwilling or unable to draw up forward-looking energy policies to ensure industrial safety and environmental protection. It is no wonder that Taiwan people were more than a bit shocked to know that the per capita carbon emissions on the island have far exceeded the quantity limits that the UN recommends. We have caused heavy carbon pollution in our surroundings, and we know nothing of it! Similarly Taiwan repeatedly suffers from shortage of water resources and electricity because of wasteful energy use, poor management and populist and emotive disputes. All this has become a worry well-known to all. In energy, environment, food, water, electricity and other areas, a great deal of waste, pollution and outdated management exists.

Human desires are insatiable. Modern technology has brought convenience to our life, but it also accelerates the consumption of our resources. Just to take the example of mobile phones, many consumers frequently replace their phones with a more fashionable model. In Taiwan, for example, the annual sales of mobile phones total 8 million but the recycling rate is only 1%. In this respect, many places are not much different from Taiwan.

Mobile phones and accessories contain valuable metals such as gold, silver and copper but also toxic substances like lead and PVC. In electroplating the casing of the phones alone, sulfuric acid and heavy metals are discharged, which then seep into rivers and the ground, and contaminate the water. If we dump the discarded phones, computers, laptops, TV sets and refrigerators in landfill or incinerate them, they can pollute the environment, with a more serious impact than nuclear waste. The reason is that before it is disposed of, nuclear waste has to go through a rigorous treatment procedure and is buried deep in the earth. Its impact on the environment is remote compared to that of non-nuclear pollutants.

In fact, "waste is a resource put in the wrong place". Discarded household appliances like mobile phones can be converted to treasures if they are put to good use. It is estimated that one kilogram of gold can be retrieved from 70,000 mobile phones and one kilogram of silver from 100,000 mobile phones. Every day, there are about 20 tons of discarded electronics accumulated in Hong Kong. How can we transform them into something valuable?

Purity Endures Like the Lotus

It takes time for nature to provide us with our limited resources. With the increasing scarcity of the planet's resources, we must learn to treasure them and make the best use of them, keeping always in our mind the human need for sustainable development.

There is something magical about recycling food waste and reusing discarded materials in such a way that something that appears to be useless is turned into something useful, but the transformation process requires the use of energy and resources, and for that reason, it is not the best solution. It is reasonable to say that environmental pollution is the byproduct of economic growth. The acts of vacuous rhetoric and politicizing every social issue not only waste tangible human resources and increase intangible stress, they also constitute a horrible social and psychological contamination.

Failing to see right from wrong, indulging in mindless boastfulness, refusing to look at things objectively — the mental impact could be more far-reaching than nuclear power and non-nuclear pollution.

"Seeing the lotus blossoms pure, I know your purity endures." These lines from a Tang poem should inspire us to look at things for what they are and not to get confused by prejudices and sensational stories. Let's start with little things to save energy and protect all things on earth.

15

In Search of Innovation in Formulating Energy Policy

The gas explosion at the Fukushima nuclear plant was caused by oxidization of the zirconium on the outer shells of the fuel rods and the resulting hydrogen that was released. This tragic accident was thought-provoking for scientists: since hydrogen is viewed as a much desired clean energy source, is there a highly efficient way to collect hydrogen and use it as an alternative energy?

Address the Energy Problem by Means of Innovation

Since the dawn of man, when our ancestors learned to drill wood to make fire and said goodbye to the days of eating the raw flesh of birds and animals, humans have discovered the use of three important energy sources — firewood, coal and oil. Before long, people came to realize that these energy sources were limited.

We also found that these energies had low conversion efficiency and that they polluted the environment, and caused people great discomfort. Nowadays, in addition to such traditional sources of energy as coal, oil, natural gas, hydropower and nuclear energies, we have started to use solar, wind, biomaterial, ocean currents and other renewable energies. These sources are so diverse as to be completely unimaginable to our ancestors.

With the rapid economic and scientific development in society, our increased demand for energy has resulted in the improvement of energy development and the expansion of exploration to places as far as the Antarctic area. We have even probed the depths of the ocean, through 3,000 meters of sea water to reach rocks on the seabed or through 1,500 meters of salt rock for the purpose of exploring oil-gas and flammable ice. Today, the exploration of new energies has become one main cause of international conflict.

In the face of a crisis, people often hope that they can explore and harness new energies. For example, as nuclear power faces new restrictions, people have begun to think about tapping geothermal sources. Japan and Iceland have respectively planned to tap geothermal sources from volcanoes.

The Fukushima nuclear accident and hydrogen explosion that followed prompted people to study how to make use of hydrogen. Since hydrogen is the main component of water, one way is e to use electrolysis to extract hydrogen from water. Some scientists have begun to study how to combine CO_2 and the excess electricity produced by solar and wind sources (which at present is not stored) to produce methane – a gaseous fuel. A related challenge is how to extract the methane stored in the frozen layers of the Arctic Ocean?

Scientific Innovation: Cornerstones of Safety and Reliability

During the Fukushima nuclear accident, a mishap in the supply of coolant water seriously damaged the nuclear power plant. Some suggested after the disaster that a reservoir should be built on the elevated land near the nuclear power plant to adequately protect the facility. Unfortunately, this theory overlooked the damage that a potential earthquake would do to the reservoir, which could be more vulnerable than the plant itself.

Employing an unreliable method to safeguard a reliable device brings more losses than gains. Unfortunately, our society has witnessed too many irrational practices of this kind. For the same reason, it will be wrong and fruitless to replace a fairly clean and reliable energy with a more harmful one. It's important to take safety and reliability into consideration while drawing up an energy policy. Let us not allow the temporary setback of the Fukushima accident to upset the need for a long-term energy strategy.

Nuclear power has become a target of public concern in many countries partly because nuclear waste produces potential environmental challenges if not treated properly. Some suggested that nuclear waste be transported into outer space by rockets. Obviously, this is an impractical proposal, as the successful launch of a rocket is even more unreliable than the disposal of nuclear waste itself. Though present-day technology doesn't allow us to completely burn up the nuclear fuel, scientists have made a major breakthrough in the recycling and treatment of the spent fuel.

If nuclear fuel could be exhausted, there would not be any nuclear waste, and the environment would not be polluted.

Allow me to cite a comparable example in our daily life. After a barbecue in the countryside, we leave when we have completely put out the burning charcoal with water. However, I think there is an even better approach. The ideal solution is to match the exact amount of charcoal to the amount of meat so that the charcoal will be completely used up when the meat is done. In this way, you don't need to use water to put out the burning charcoal as there is no burning charcoal left behind. Such a method will not only ensure environmental protection and safety, but also avoid wasting energy.

The safety of a nuclear power plant depends on three vital factors. These are high-quality equipment, well-trained personnel and a strict management system. This is why all power plants under development must take all potential accidents into consideration, including possible mistakes the operators could make and the common fallouts brought on by various natural calamities and man-made disasters.

Stringent quality control measure and transparency are crucial to effective management. In August 2013, South Korea faced an unprecedented power shortage after the government's decision in May to keep four nuclear reactors offline because of forged safety certificates for components used at the facilities. Earlier in November 2012, the government had already shut down three of the country's 23 reactors due to similar scandals.

The shutdowns have resulted in a potential energy crisis in the country, provoking strong complaints against the government's failure to meet electricity demand in the hot summer. The scandals could also hurt South Korea's efforts to export its nuclear power technology to other countries. On the positive side, this could compel the government to instigate more reliable quality control and monitoring measures to ensure safety of power plant operations.

The above-mentioned discussions and examples may involve only a few concepts, but they serve to illustrate a point, namely, that innovation has the power to turn the rotten into the miraculous and deliver us from awful predicaments and inevitable limitations. In short, while formulating a policy of science and technology, we must take into account innovation and environmental protection and, in particular, adhere to the principles of safety and reliability as the cornerstones for its implementation.

The Myth of Building a Nuclear Power Plant by the Sea

After the nuclear accident in Japan, there emerged myths about building a nuclear power plant by the sea, especially those similar to the Fukushima Daiichi Nuclear Power Plant. Obviously, no nuclear power plant has developed any adequate contingency plans to deal with 16- to 17-meter high mountainous waves. Meanwhile, no research has ever been conducted on the situation where sea water has to be poured into nuclear reactors. Nor is there any record of such a scenario. Therefore the Japanese were in helpless confusion when the accident occurred. Moreover, hydrogen explosion was never a focus of attention in the past.

If a hydrogen-releasing device is set up on the rooftop of a nuclear power plant in the future, even if a disaster similar to the Fukushima earthquake were to occur, the safety of the plant would be enhanced.

That no precautions were in place for the eventuality that charging cables might be inundated by huge waves is inconceivable. For that matter, neither were there any back-up cables. It seems that management negligence is to blame. The "*WASH1400*" report lists in detail all kinds of risks involving nuclear power generation. However, the fault tree study in "*WASH1400*" contains no suggestions for the kinds of problems that the Fukushima Daiichi Nuclear Power Plant suffered. In addition, could there be other problems that we are not yet aware of?

All this has highlighted the importance of scientific research. Unfortunately, our research is far from being solid.

Formulation of an Optimal Energy Policy

Energy sources and environmental protection are the problems of the day. They are also topics to which we have to pay special attention when we consider sustainable development. As far as energy sources are concerned, we have many choices, such as water, coal, natural gas, solar energy and nuclear energy. Whether we select one energy or a combination of several energies, we must take heed of the following three factors: 1) daily necessity; 2) reliability, safety and sustainability; and 3) social well-being and economic growth.

In each of the different kinds of energies available to us, there is much room for improvemei.t in raising its efficiency. For example, according to a report on the "China Energy Conservation and Supervision Website" (www.cecs.gov.cn), the unit output ratio of mineral resources in China is 1/6 of that in Germany, 1/10 of that in the US and 1/20 of that in Japan. The output ratio of one ton of standard coal in China is 28.8% of that in the US, 16.8% of that in European Union and 10.3% of that in Japan. The average output value of one-cubic-meter of water in the world is $37. In the UK, it is $93, while in Japan it is $55, and in Germany it is $51. In China it is only $2. It is not acceptable to use low-efficiency energies; nor is it acceptable to tolerate the low-efficiency use of any energy.

With quality of life foremost on their minds, people cannot make their choice of energies without taking into consideration reliability, safety or sustainability.

In order to improve our living standard, the relationship between our increased demand for energy and the price paid for the use of energy must be considered, as described in Chapter 14. This is the only way for us to maximize social well-being with limited resources.

We must keep a balance between these three factors and not just focus on one of these factors to the exclusion of others. Unfortunately, many countries and regions do not have an energy policy ready for implementation. It is little wonder, then, that Taiwan is still hesitating over whether to complete the No. 4 Nuclear Power Plant in Lungmen even though the Taiwanese have already invested more than $10 billion in the project in the past 20 years.

Energy Conservation: Only a Precondition for the Formulation of an Energy Policy

Nowadays, any energy policy must be discussed on the basis of energy conservation which, however, is only a necessary and a prerequisite to any selection of energy sources. Energy conservation and energy development are not mutually exclusive, and energy conservation is not sufficient for energy development.

It is misleading that if we save energy use to the extent of the current energy consumption produced by the nuclear power, then we would be able to remove the use of nuclear energy. Energy conservation has very little to do with the use of nuclear energy unless we do not use any energy at all in our daily life.

Therefore, our energy policy should be an optimal combination of different energies with energy conservation and carbon emission reduction as its prerequisites. The production of green energies can only be achieved after we draw up and implement a successful energy policy. At the same time, energy exploration must remain environmentally friendly.

However, "a nuclear-free homestead" should not be the assumed ultimate goal when we draft an energy policy. If we look upon this as the ultimate goal, we will be compelled to rely on coal, natural gas and other fairly polluting energies. As a result, it will be detrimental to the well-being of the general public and impair their quality of life. What's more, the energies we use are not reliable, safe or sustainable. A totally green energy or a nuclear-free homestead may be feasible for some countries, and is too expensive for many others with insufficient natural resources to afford.

It follows then that nuclear energy is still the optimal choice for the majority of countries and regions including those in East Asia. However, due to the influence of populist politics on the one hand and people's concerns about its safety on the other, some societies find it difficult to accept nuclear energy.

In order to draw up a methodical energy policy and put it into practice, it is necessary for the government to increase transparency when announcing the operation of a nuclear power plant and ensure that an independently operated nuclear supervising agency is in place. To help ensure this, an independent global fund could be entrusted with the tasks of developing competence and skills, and training professional personnel to promote and organize open discussions to analyze and debate energy issues and new developments.

Smart Grid

Nowadays, the global demand for electricity is increasing continuously. But the present-day transmission system fails to satisfy the growing demand due to its uneven distribution and poor supply quality. Power disruptions and restrictions on power supply are common occurrences in many places.

It is not unusual to see the following awkward situation: power shortages in some places and power surplus in others happening simultaneously. Even in the same place, there are shortages of electric power at certain times and surpluses at others. Especially in a time that sees more and more energy shortages, such inconsistencies are especially distressing. It should be an urgent task to balance the supply and demand of electric power and control the costs of energy. This is also true for food supply and food consumption; for a great part, it is a distribution problem (see Chapter 14).

At the same time, people are worried about environmental pollution caused by the burning of petrochemical fuels and coal. They demand that the government develop wind, solar, bioenergy and other renewable energies. In combination with a broader spectrum of energies such as hydropower and nuclear power, we can find the optimal way to supply electric power.

In addition, we should develop a precise control system and make full use of high-speed computers and advanced electronic communication technology to work out an optimal strategy for guiding power supply. Measures should be adopted to ensure easy connection, storage, distribution and electricity conservation.

While the general public has stepped up their demands for highly reliable, efficient and sustainable electric power, guarding against terrorist attacks remains a challenge. All these factors have combined to introduce the smart grid, a robust power generating, transmission and distribution network.

The environmentally friendly smart grid is an advanced platform that disperses power risks, complies with the market demand and sets a reasonable price for electricity. By applying the remote control function of a high-speed computer and real-time automatic distribution, the smart grid will provide the 21st-century metropolis with a practical and reliable system of electrical load, energy storage and power transmission.

The smart grid will play an important role in energy conservation and the increase in energy efficiency. This innovative contribution was made possible by such high technologies as computer and auto-control in the field of environmental protection. The role the smart grid will play in optimizing the distribution of a spectrum of energies and utilizing electric power requires further promotion and intensive study.

Democracy, Populism and Innovation

After the Fukushima nuclear accident, the reliability of energy became the focus of attention. I was invited to make a keynote speech on the design of nuclear safety at the *2011 Annual European Safety and Reliability Conference* held at Troyes, France, in September 2011. I talked about environmental protection, as well as energy and industrial safety in this region. Progress has been made in innovation and management in Taiwan, but its poor performance in practice needs to be addressed. The unreasonable electricity price, for instance, is a typical example.

While Hong Kong is a lively and prosperous city, there is a growing polarization between the rich and the poor. The government is conservative when drawing up policies, talking big but doing nothing. It must get rid of its passive and inert working style and shoulder the task of drawing up an effective science and technology policy, focusing especially on the implementation of responsible energy use.

We must select appropriately from among a spectrum of energies both today and in the foreseeable future. Firstly, in my opinion, we must reduce coal-generated power because of the pollution and many casualties.

When drafting a policy, we must weigh the pros and cons of energy technologies and clearly define the relationship between the growth of science and technology, sustainable development and the improvement of the citizenry's well-being. Publicizing the concept of reliability will solve many social problems. We should judge everything on its own merits and not by the political fallout that it may trigger. This principle should apply not only to the upgrading of a power plant, but also to the improvement of high-speed train, the promotion of environmental protection, the enhancement of higher education and the building of a democratic society.

Our long-standing conservative culture in many places in Asia lacks the spirit of innovation. Faced with the prospect of reform, some people will always obstruct innovation, because reform will affect their vested interests. If the benefits of the reform are not immediately obvious the reform will fail to win the fervent support of the populace. What's more, most people like to adopt an aloof attitude and remain silent. "Do nothing; make no mistake" may be a safe strategy, but when it comes to energy safety, it is not enough to set one's mind at ease.

The progress of democracy in many places including Taiwan and Hong Kong is still in the so-called infant mortality stage (See Appendix I) at which populism is the most prevailing practice. A superficial kind of democracy may be able to redistribute wealth to some extent in the short term, but only the more substantive kind of democracy will, through supporting innovation, go beyond mere redistribution to widen and deepen the foundations of the wealth of a society. Unfortunately, democracy in our society has been seriously misused and misguided. The understanding of the true nature of democracy depends on our constant examination, analysis and study.

Without innovative science and technology policy, we will have no basis for addressing energy and environmental protection concerns.

16

Practice Makes Great

Steve Jobs, the co-founder of Apple Inc., is an embodiment of innovation. He established a new notion of computer use and Apple's products, be they the iPhone, iPad or others, are well-known examples of "thinking outside the box". The fertile soil from which Jobs' creativity grew encourages innovation and invention, coupled with a realistic market orientation. Besides innovation, the key to Jobs' success lay in his desire for perfection in the making of his products. As far as that is concerned, he was above all a practitioner of ideas through and through.

There is a curious phenomenon in Chinese society in putting too much stock in a person's degree and placing undue emphasis on the transmission of book learning. Meanwhile, too many people worship elite schools, not realizing that education credentials are not necessarily a free pass to success. After all, a university degree is only a certificate of one's qualifications. It merely shows on paper one's exam scores at a certain stage of one's education. It does not guarantee a good job.

To be sure, ordinary mortals that we are, a degree certificate may help in some ways. A respectable degree may bring with it an opportunity to get an interview, but gives no assurance that one will be hired. In the workplace, a degree is not as important as real learning, which, in turn, is not as important as practical experience. The degree one gains in one's earlier years does not play a major role in later achievements. To get hung up over a degree from a prestigious school will only hold back the full flowering of a person's potentials and be of little value to society.

PhD Stands for… What?

Many people blame bad luck for not realizing their potentials and often end up wallowing in self-pity, attributing their failure to the fact that their degrees are not recognized. This makes no sense. We often think too much about the PhD degree. Ironically, the Chinese word for PhD has often been carelessly written to resemble a word that could mean "a dummy". There seems to be quite a few PhDs of this kind among us.

Here is a story about a crab and a bat. Neither of them is happy with their lot; for the bat, it is because he sleeps hanging upside down, and for the crab, it walks sideways. One day, they decide to study for a doctoral degree together. And after they receive their PhD degrees, the bat continues to sleep upside down and the crab continues to walk sideways. The degree has done nothing to alter their nature. Worse, in the case of some people, a PhD degree seems to make them a dummy.

To succeed in a career requires discipline, passion, perseverance, and an unswerving focus on one's study and work. Steve Jobs of Apple, Bill Gates of Microsoft, Mark Zuckerberg of Facebook and Michael Dell of Dell Inc. did not need a degree to bring millions of degree-pursuers under the spell of their high-tech products.

According to data in 2005, most CEOs in Fortune 500 companies started their careers from the bottom and gradually rose to be company leaders, attesting to the fact that dedication to one's work is very important. This runs contrary to popular belief. Moreover, to be a CEO, one does not need to have an Ivy League degree. As a matter of fact, more graduates of American public institutions become CEOs, again belying popular belief. More than a third of Fortune 500 CEOs are just bachelor's degree-holders, with engineering degrees, and not business, being the most common among them. MBAs, EMBAs, and PhDs are not necessary or essential.

The current situation in Taiwan and Hong Kong is exactly the opposite. The job markets in these two regions have been swarmed in recent years with PhD degree holders looking for jobs more suited for a master's degree-holder, and people with a master's degree looking for jobs more suited for those with a bachelor's degree. The value of degrees is depreciating rapidly.

Another common phenomenon is that there is a high percentage of CEOs in the US who have served in the military and had international experiences. Around 15% of Fortune 500 CEOs have been in the army, a rate that is far higher than the rest of the American population, and 33% of them have had international work experience, demonstrating once again how crucial internationalization is.

In addition to over-emphasizing the value of a degree and going after prestigious schools, Chinese society today appears to encourage people to take the short cut of outsourcing or sub-contracting all manufacturing work as a fast way to get rich. This explains why we will never be able to create an Apple Inc. for Taiwan or Hong Kong. For the same reason, our local talent is greatly limited by the sheer lack of practical industrial experience. They become trapped in our "pond-style" economy. They are handicapped by the lack of practical experience to be able to participate meaningfully and effectively in discussions and formulations of policies and legislations which impact occupational safety or environmental protection issues in our society.

According to Taiwan's Ministry of Education, the island is one of the regions that awards the highest number of bachelor's degree, master's degree and doctor's degree. In mainland China, there were about 6 million college graduates in 2010 alone. The reason that we fail to approach issues such as occupational safety and environmental protection rationally obviously does not lie in the fact that we lack learned scholars or experts or leaders with advanced degrees.

The more likely answer is that our society is flooded with emotional slogans, while there is not enough practical spirit or resolution to carry out what is right. Though society is glutted with people with advanced degrees, occupational safety and environmental protection issues do not get resolved. Sometimes, because these people are equipped only with their degrees but in other ways refuse to submit to reason, they constitute a force of resistance to environmental protection, showing once again that learning is more important than a degree, and experience is more important than learning.

Putting Knowledge and Practice Together

To protect the environment and improve occupational safety, we need to undertake a comprehensive review and have systematic planning, in accordance with Jobs' way of pursuing perfection. This is the right way to go.

Here again I want to talk about the pursuit of perfection. AT&T in the US, the father of the modern telephone as well as the pioneer of quality control, dominated the global telecom industry for three quarters of a century. Bell Labs, one of its research units where I once worked as a young scientist, was very unique and had the world under its feet for a while.

But since the 1990s, AT&T has been in decline and has been superseded successively by Motorola in the US, Ericsson in Sweden, Blackberry in Canada and Nokia in Finland. However, the dominance of all these brands was also not for long. They were all later beaten by Apple Inc. led by Jobs. Meanwhile, less than one year after Steve Jobs' death, the crowning position of Apple's iPhones in the market was taken by South Korea's Samsung in May 2012. Who would have thought that possible?

Samsung's success has behind it the accumulated experience of two to three decades of practice and effort, and is attributable to its pursuit of perfection that puts knowledge and practice together. But believe or not, Samsung will also fall behind or be replaced in the future if it doesn't innovate or keep improving in its pursuit of excellence.

Ah Q's Daydream

Ah Q, who is a worrywart for the future, has spent some time studying statistics, but he never fastens his seat belt when he is in a car or washes his hands before he eats or after he uses the toilet. In Ah Q's house cockroaches and rats are running amok. He fears the dark and ghosts so he turns on all the lights in his rooms.

In order to utilize marsh gas to generate electricity on the one hand and save water on the other, he has given up flushing the toilet and uses a traditional latrine instead.

Ah Q drives a deluxe car. To boost the energy efficiency of the car, he never turns on the air-conditioner even in the sultry summer. What's more, he never allows passengers to open the car windows and instead provides them with just a folding fan.

While walking on the streets, he always keeps a distance from wire poles for fear that the electromagnetic waves might undermine his health.

To me, Ah Q has studied statistics in vain. An ancient poem reads, "If only I could lay my hands on the famous Zhongshan brew /to get myself drunk for 1,000 days until peace reigns throughout the world!"

If we do not weigh the pros and cons of different types of energy and disregard the imminent danger of industrial safety and the long-term consequences of environmental pollution or take lightly the results of natural calamities or man-made disasters, we are merely daydreaming and looking far and wide for a solution that may lie close at hand. Poor Ah Q, where would he find the Zhongshan wine that would keep him in his oblivion until peace comes?

In order to respond to changes in the world and the development of human history, we must abandon the Ah-Q mentality and populist appeals in society. For the purpose of promoting social well-being, we should adopt a rational and analytical approach in understanding the essence of a spectrum of energies and environmental protection. As a conclusion to this chapter, I quote Cheng Hao's poem, "A Poem Written on the Wall of the Huainan Temple":

> *On my way from north to south I feel free to take a rest any time,*
> *The autumn wind sweeps away the white-flowered weeds on the river.*
> *I'm not a passing traveler sentimental about the melancholy season,*
> *Let the mountains on both sides of the Chu River grieve for each other.*

Appendix I

The following is a revised version of my keynote speech delivered at the 11th Annual Conference of China Science and Technology Association in Chongqing on September 8, 2009 attended by 7,000 participants. The English version has been published in the February 2011 issue of *IEEE Reliability Society Newsletter*.

Reliability Through the Ages

Foreword

Reliability and maintainability are having an increasingly significant impact on people around the world. As a matter of fact, the concept of reliability originated from its application in military affairs. As early as before World War II, the US started to attach great importance to research on reliability in order to meet military requirements. In addition, reliability has played a pivotal role in NASA, the application of nuclear energy and in the operation of man-made satellites. Human civilization was, sometimes, derived from wars.

The conclusion reached by a philosopher—"The civilization we enjoy today originated from quarrels and wars"—may court dispute, but so far as reliability is concerned, his remark is an undeniable truth.

Reliability is a way of evaluation. Judged by reliability, no system is permanent in history. This article is devoted to the illustration of the close relationship between reliability and human beings from the perspectives of engineering history and human society, including urban architecture, aviation, auto manufacturing and insurance. In addition, the cure rate for diseases, survival rates after surgery, the warranty period for automobiles, the maintainability of aerospace and nuclear energy, and value-preserving rates in the insurance industry are all manifestations of reliability.

Of all the systems, human beings are the least reliable, to be followed in succession by casualties caused by traffic accidents and wars. The failure rates of microelectronic systems are fairly low. Compared with all the above-mentioned systems, the accident rates for airplanes and nuclear power plants are even lower.

In this appendix some pertinent solutions to various reliability-related problems are offered to help human beings reach the state of "enslaving rather than being enslaved in material applications".

The Essence of Reliability

The role played by reliability and maintainability in traditional industries and modern technical products is such a broad subject that one article cannot cover everything. This article is intended as a discussion from a historical perspective and takes a panoramic view of the relationship between our daily lives and reliability in the past, present and future.

Though the study of reliability started only after World War II, its findings have exerted great influence on us over the past decades. For example, the cure rate for diseases, mortality rate for cancer, the effects of some medications on human health, and the harmful effects of residual pesticide on farm produce are all related to reliability.

Because expected life is mentioned in healthcare, the reliability that we are discussing is probability. Probability is usually beyond the comprehension of the lay reader. For example, a patient is told after carcinectomy that the survival rate is 90%. This is a vague expression. An easier concept to comprehend is how long one is expected to live.

Expected life, which is closely related to probability, is in fact another indication of reliability. Expected life is relevant to a warranty period in the manufacturing industry. In the early days, the US auto industry offered a one-year warranty to customers. Later, Chrysler changed the one-year warranty to a three-year warranty. Toyota then extended its warranty period to five years and Chrysler provided a seven-year warranty. Recently some auto manufacturers have started to offer a life-long warranty for certain main parts in cars.

Did the auto manufacturers lose money by issuing such a long warranty? The answer is no. If so, car manufacturers would no longer be in business. All in all, the auto manufacturing industry remains very profitable. The reason that the auto manufacturers could still make a profit by the end of the 20th century lies in a thorough study of the quality and reliability of their products, a careful analysis of the expected life of their products, and the cost of the expected life. The improvement of reliability benefits both the manufacturers and the customers who use the products.

In addition, reliability covers the design of risk analysis, the fault tree analysis of nuclear energy and the fault tolerant analysis of software. For example, the insurance industry is a continuation of reliability. The fundamental aim of the insurance industry is to earn money from the insured party, and they will not do unprofitable business.

To attend college today is much more complicated than before. Formerly a candidate only needed to fill in an application form. Nowadays the multiple-enrolment system makes the process so complicated that applicants and their parents hardly comprehend it. The complexity of choosing one's ideal programs and specialized subjects poses some knotty problems. This belongs to the category of reliability, too. That's why education is also related to reliability. The synonyms of reliability used in various industries are listed in Table 9.

Politicians often make irresponsible promises or put up empty slogans. We may as well listen to the figures listed by politicians, but we are not supposed to believe them because of great uncertainty about these figures. Why are the remarks made by politicians so unreliable? Why do politicians elected by the majority utter so many high-sounding words? Why are there so many people who still believe their slogans?

Reliability Bottleneck and Historical and Societal Track

The above-mentioned general survey and essence of reliability can be summed up in Table 9. What is reliability then?

Table 9 Synonyms of reliability in various industries.

Industries	Synonyms of reliability
Agriculture, Pharmaceuticals	Cure rate (Survival rate)
Medicine	Expected life
Manufacturing	Warranty period
Aviation, Space, Nuclear energy	Maintainability, half-life
Insurance	Hedge ratio
Education	Admission (Graduation) rate
Politics, Economy	Policy stability

Take a story about Confucius for example. Confucius had 3,000 disciples. Of them only 72 were top students who were as qualified as today's academicians. Of the 72 disciples, Yan Hui was the most outstanding one. Confucius said, "For three months at a time Yan Hui does not lapse from benevolence in his heart." Such a talented student could only be free from lapsing for three months. One month consists of 30 days and three months consist of 90 days. Calculated by days, his accident rate is about 1%. What about the others? There are changes every day and even every hour. The comment made by Confucius about Yan Hui can be considered as a good example of the best performance of a human being's reliability.

Reliability involves quite a lot of factors, of which "human being" is a very critical bottleneck. As we know, Yan Hui's reliability is in the upper bound — the maximum of human performance. There are a lot of subsystems under the main system and even more unreliable factors.

Why do we hold the opinion that reliability is worth studying from the historical perspective and is one of the priority subjects for the future development of science and technology? Take Li Bai's famous poem "A Trip to Kiang-ling" for example. His last line "The skiff slid myriad mount ranges away" enjoys great popularity. Li Bai was a native of Sichuan province and described the Three Gorges of the Yangtze River in some of his poems collected in *Three Hundred Poems of the Tang Dynasty*. In "A Trip to Kiang-ling" he described the Qutang Gorge—one of the Three Gorges, which is close by to Fengjie. I'd like to tell you a related story about White King.

Wang Mang was a scholar living in the last years of the Western Han Dynasty. He usurped the throne for the sole purpose of governing the Western Han Dynasty with Confucian doctrines. He advocated a lot of good administrative policies, but failed to carry them out due to his poor execution and considerable resistance. Gongsun Shu, a high-ranking general under his command, crowned himself king on the slope of a hill in Fengjie. He was known as White King. After Wang Mang's downfall, White King was the only surviving successful ruler. Even though Gongsun Shu was defeated by Liu Xiu not long after that, the local people built a temple to show their gratitude to him. The temple was called White King Temple and the hill was thereafter called White King Hill.

According to Chinese feng shui theory, White King Temple, sitting against a hill with water on three sides, is located in the most favorable surroundings. As a result, many collections of stone tablets were established in White King Temple, and the statues of Guan Yunchang (a household name of a general serving under the warlord Liu Bei in the late Eastern Han Dynasty of China) and Liu Bei were placed there.

White King Temple became a tourist attraction. The local people in Fengjie thought White King Temple would last forever.

According to reliability theory, nothing will last forever. As a matter of fact, after the Three Gorges Dam was built, the White King Temple area was inundated and became an island. Only the temple stands above the water. No one can ensure the eternity of such a fabulous and famous historic site. Isn't this a phenomenon of reliability's influence on a city?

Reliability is different from quality. The most noticeable difference between reliability and quality is that reliability is a time function whereas quality is not. As a time function, there is no constant phenomenon.

Take the well-known cliff-side paths of Sichuan for example. The paths are 65 kilometers long. Whenever the Yangtze River flooded, navigation became impossible. That's why people dug out cliff-side paths during the past 1,500 years. The distinctive cliff-side paths, which were used to facilitate transportation during the Anti-Japanese War, have a historical significance. They are now submerged due to the building of the Three Gorges Dam. Nobody had ever predicted this. Even the cliff-side paths are not independent of the control of time factors.

While we allow some historic sites to be submerged, other historic sites have been divulged, such as the terra cotta warriors and horses from the Qin Dynasty. Even the First Emperor of the Qin Dynasty would never have expected that his mausoleum would have been unearthed. If we look at the world from this perspective, the Three Gorges and the cliff-side paths of Sichuan may re-emerge in 2,000 years. The terra cotta warriors and horses may be buried again.

Nothing is permanent or everlasting in this world. In the study of reliability we should take heed of this fundamental principle.

Black-dress Lane: An Embodiment of Reliability

The Tang Dynasty poet Liu Yuxi's famous poem "The Black-dress Lane" demonstrates the intangible impact of reliability on the changes of a city.

During the period of the Three Kingdoms, the Kingdom of Shu was first defeated by Wu. Wu was situated south of the Yangtze River. One year after Wu defeated Shu, Wu was defeated by Wei, hence the end of the period of the Three Kingdoms. After Wu's downfall, Sima Yan usurped the throne of Wei and crowned himself the Wu Emperor of Western Jin Dynasty. He made Luoyang the capital. After he founded the Western Jin Dynasty, Sima Yan wanted to earn popular support. Misguided by many scholar-officials and hindered by the internal turmoil created by eight princes, he failed to invigorate his country. Instead, national power declined. Not long after the northern tribes moved southward.

After a period of turmoil brought about by the five tribes, the Western Jin Dynasty was eliminated. That is why the Western Jin Dynasty is considered a defeated kingdom. Content to exercise control over part of the country, its successors moved the capital to Jiankang (Nanjing today) and founded the Eastern Jin Dynasty.

How are these historical facts pertinent to the poem "The Black-dress Lane"? At that time Nanjing was controlled by the two most influential families of Wang Dao and Xie An. It was picturesque near Red Bird Bridge. Beautiful flowers and a constant stream of horses and carriages made Black-dress Lane a prosperous and flourishing place.

When Liu Yuxi visited Red Bird Bridge 500 years later, the prosperous Black-dress Lane was nowhere to be found. That's why he wrote the following poem:

> *The wild flowers spread near the Red-Bird-Bridge side,*
> *The setting sun was passing the Black-dress Lane by;*
> *Swallows once nestled in the high official chamber,*
> *Now flying into the plain house of the commoner.*

(Translation by Manfield Zhu, with adaptation)

That means the expected life of a city is only 500 years. Everything has a lifespan. Nanjing was not the only city that follows the law of expected life of a city; other examples include: Athens, Greece of the fifth century, Kaifeng, China of the 11th century, and Florence, Italy of the Renaissance which all have faded in less than 500 years from their peak time.

A politician in power may boast a life of splendor, but his days of glory are numbered. Judging the world from this perspective, few big cities can remain prosperous for over 500 years. Liu Yuxi can be regarded as a person of foresight because he was able to predict that the optimum lifespan for a city was 500 years.

Public Construction from the Perspective of Reliability

The five-star Hyatt Regency Hotel in Kansas City collapsed suddenly 20 years ago. Not long after, the Hyatt Regency Hotel in Manila collapsed, too. That means something must have gone wrong with the design.

When the construction of TAIPEI 101 (Taipei Financial Center) was completed, I wondered how long before something went wrong. I was worried because I knew the quality of a public construction depended on simulated tests. However, Taiwan does not usually pay enough attention to the quality of public constructions. Just as I expected, not long after it was constructed, an iron plate was blown off by a gust of wind and many passers-by were hurt.

Reliability is of primary importance in the construction of a power plant. A short-out in a well-known US power plant one night in 2003 caused a blackout in the border area between Ohio and Michigan. The power failure had a huge impact on society. We can conclude that reliability is essential to power transmission and nuclear power plants. If we experience a sudden blackout, transport will come to a total standstill and traffic will experience gridlock. All of us will feel at a loss in the streets. Can society cope with such chaos? Do you think people would rush into the streets in an orderly fashion?

I cite these simple examples to show that reliability has a great impact on our life.

The airplane is a fairly reliable means of transport, but there are still a lot of airplane accidents, and they can cause heavy casualties. A few years ago, a door of an airplane over Africa suddenly opened mid-air and many passengers were sucked out of the cabin. It so happened that someone on the ground took a photo of the incident. Just as I mentioned before, reliability is a time function. Inevitably accidents will occur in due course, sooner or later.

That is why it is very important to study reliability. When shall we stop the lifespan of a system? When shall we make a simple and correct estimate about the costs involved in the lifespan of a system? For example, space research in the US is undoubtedly first-class, but there have still been several accidents, often caused by minor problems but resulting in huge losses. Every time there was a spacecraft accident, NASA would ask me for help. Unfortunately, as often as not "it's too late"!

Many corporations and enterprises have ignored the study of reliability. They tend to adopt remedial measures hurriedly only after an accident. If they had taken preventive measures, they could have reduced their costs. All the above-mentioned examples indicate that reliability will have a great impact on our cities, bridges, historic sites and on water and soil preservation.

Another illustration is the story of a great dam in the US which burst after a small crack gradually grew bigger resulting in the flooding of an entire city. How long did it take for a crack to expand and the leaking water to breach the dam? Just six hours. As an old Chinese saying goes, "one ant hole may collapse a five-kilometer dyke". This is why we have courses on reliability and related research is conducted in departments of nuclear engineering, hydraulics and civil engineering.

Accidents in nuclear power plants are naturally very worrying. One occurred in the Three Mile Island Nuclear Power Plant in the US. Thanks to well-conceived designs, there were no serious health effects. But the accident at the Chernobyl nuclear plant in the former Soviet Union was devastating.

It is interesting to remember that both the US and Soviet accidents took place in the middle of the night. It is worth recalling what Confucius once said: "Human beings are the least reliable". A more efficient way to improve the well-being of mankind would be to improve the human factor, have politicians talk less about figures, and see the reliability of human beings increase from 0.1 to 0.15. This would be 10 million times more efficient than simply investing in high technology.

How Reliable Are High-tech Products?

About 50 years ago the integrated circuit was composed of only four electronic transistors. People tended to ask its inventor, "Will the integrated circuit products break down?" The inventor would answer, "They will not break down. They will work for ever." How would that be possible? The greatest challenge that electronic products have today is not how to make 8- or 12-inch wafers, but how to face the challenge of reliability.

In present-day electronic factories, reliability is considered their top secret. I once took my students to the factory of Texas Instruments (TI) to visit their reliability-related facilities. Because one of my students used to work at a Samsung electronics factory in South Korea, the personnel at TI told me that the reliability of their electronic products was their top secret and that they would have to turn down our request because of the student's previous connection to Samsung.

Why are the reliability-related documents rated as top secret by electronic product manufacturers? Firstly, the acceptance rate of electronic products is very low. Secondly, they do not want to reveal that they are turning out unreliable products. That is why no US magazine has ever published the acceptance rate for electronic transistors produced in US factories. Not once! The reason is that they do not want people to know that the acceptance rate of their electronic products is actually very low.

In addition, the reliability of initial products in today's nanotechnology industry is similar to that of human beings, i.e. less than 10%. If somebody can provide data about reliability or makes some proposals to improve reliability, nanotechnology will certainly have a bright future.

Some people suggest that the lifespan of human beings is relatively long while that for nanotechnology is short. A new product will appear before the lifespan of the previous product is completed. Does this mean that reliability is not so important? On the contrary, as the lifespan of a product is short, demand for higher levels of reliability must be satisfied. Simply because we have to use a new product even before we fully understand the reliability of the previous product, we must not only control and manage the reliability of a new product, but also design its reliability.

In respect of its design, reliability covers a wide range. I mentioned ordinary life and early life. Let's look at aging. Take the graying of hair for example. The location where every one's hair turns gray is a random distribution. Even twins' hair may turn gray in different areas of their heads.

Stress, Strength and Aging

Recent research in the US cited one of the 10 biggest challenges in science as: How long can a human being live? The answer is: human beings' average life expectancy will be 120 years by the year of 2050. This is a random phenomenon for reference only. So far as height and life expectancy are concerned, the younger generation has surpassed the older generation. This phenomenon reminds me of the relationship between stress and strength.

As shown in Figure 7, stress and strength are mechanical functions. What strength can we adopt to resist outside stress? If stress exceeds strength, we will collapse. For example, nowadays girls like tall handsome

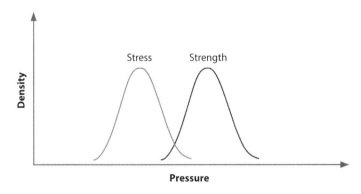

Figure 7 Stress and strength.

boys. That's why the present-day younger generation is taller than before. A young man with a height of 173 centimeters used to stand at the seventh or eighth position in a line of 10 soldiers in Taiwan 30 years ago. Today, 173 centimeters is the men's average height in Taiwan. That means men with a height of 173 centimeters in Taiwan who belonged in the top 25% 30 years ago are now in the top 50%.

Does this mean that the taller a person is, the more society will benefit? That's not the case. Tall men are also at a disadvantage. For example, they bring about more environmental pollution and have an unfavorable effect on materials around them. Do you believe me?

When taking a flight, a tall man will feel uncomfortable because he can't stretch his legs. With the increase in height comes an increase in weight.

According to my calculations, if people's average weight increases from 150 pounds to 200 pounds, an airplane with a seating capacity for 45 would be able to accommodate only 37 passengers safely. If the average weight reaches 300 pounds, the plane can only take 28 passengers. Once there is stress, there will be proper strength to resist it. These two factors are relative. If we take the time function into consideration, this relationship will become more complicated. The greater the disparity between strength and stress, the higher the reliability will be. The parts which do not overlap will never be zero because these two functions are both continuous functions of time.

Why do I mention this problem? I said before that reliability is a time function and that it is the result of a contention between strength and stress. When stress exceeds strength, accidents tend to occur. When stress is far inferior to strength, accidents will be less likely.

Infant Mortality

The "infant mortality" of reliability plays a very interesting and important role today. People tend to change their computers and cell phones every two years. That's an issue related to infant mortality. Infant mortality originally means within 72 hours after its birth. As for an electronic product, it means within one year after its manufacturing.

A great challenge facing us is how to reduce infant mortality before the product is delivered to the customer. The reliability-related problem for software mainly occurs during the period of infant mortality.

The manufacturers offer a warranty period for their products so as to ensure that they will provide free maintenance and replacement if the products break down during their lifespan. Have you heard about a warranty issued for software? Who can guarantee that their software will never break down? I haven't seen such a warranty. On the contrary, I only saw a notice on the package of some software: "The durability of this software is not guaranteed." As the period of infant mortality of software is especially long, the study of its infant mortality is highly important.

When SARS was rampant in 2003, newspapers said at the beginning that "the mortality rate of SARS is very low". Take the following figures for example: 100 patients still hospitalized, 21 discharged and four dead. Mortality rate can be calculated in various ways.

The first way is: the mortality rate is 4/(100+21+4) = 3.2%. The initial report about its mortality rate was carried in the newspaper. This was based on the hypothesis that none of the 100 hospitalized patients would die. That was actually the lower bound of the mortality rate of SARS. Anybody who has some knowledge of biostatistics knows that it was only a rough calculation.

The next day another improved calculation was published in the newspaper, i.e., 4/(21+4) = 16%. This way of calculation excluded the hospitalized patients as if they were not patients at all. That ought to be the upper bound of the mortality rate of SARS. This was not a correct calculation, either.

The correct answer should be between the lower and the upper bounds (See Table 10).

How to calculate the mortality rate of SARS remains a complicated problem. Many doctoral dissertations have been devoted to this subject. Thus we are faced with another problem: "Should we serve reliability or should reliability serve us?" Professors and students in colleges often publish theses. Do we write theses for fun? Do we wish to serve society with the results of our research published in our theses?

There is another way of evaluating the mortality rate of SARS, introduced by an article published in a newspaper. This article divided people

Table 10 Different calculations of mortality rate from SARS.

Ways of calculation	Mortality rate (supposing 100 patients still hospitalized, 21 discharged and 4 dead)
First	4 / (100 + 21 + 4) = 3.2%
Second	4 / (21 + 4) = 16%
Third	3.2% < ? < 16%

into three age groups: below 24, between 25 and 60, and above 61. There were 55, 205, and 40 deaths respectively in these three age groups. The data was collected during a certain period after SARS had spread out. The newspaper reached the conclusion that "people between the age of 25 and 60 are likely to die."

A few days later this conclusion was corrected, because in Taiwan's population structure, the ratio of these three age groups is 2.5:6:1.5. The age group between 25 and 60 is the largest. When you calculate mortality rate, you have to take population ratio into consideration, too.

So far as population mobility is concerned, the young people are the most active. The mobility ratio of three age groups is 3:5:2, and the second age group surpasses the other two. But the alertness of this age group is the lowest.

Without the addition of the other factors the mortality rate of the second age group is the highest. However, when we take additional factors into consideration, the mortality rate of the old people is the highest while that of the young people is the lowest.

If we don't take all the related factors into consideration, I believe the mortality rate of SARS among people in mental hospitals is even lower. Because they have fewer visitors and have the least access to the outside world, they are not likely to contract SARS. This conclusion is, of course, absurd. This is a simple calculation.

In the field of biostatistics the survivability of SARS patients is reliability.

Let's take the launch of a spacecraft for example. Analysis shows that flight vehicles, whether aircraft or spacecraft, are at their most unsafe when take off or landing. The airline company or the space administration is unable to ensure their safety. As a matter of fact, the space center is unable to ensure the safety of any space flight. As a result, reliability is not only important, but also incomprehensible and uncertain. We must exercise extreme caution in this respect.

A similar case is the assembly of parts of equipment and the initial stage of a new company, both of which are extremely unreliable due to the function of "infant mortality".

Nuclear Energy Industry

From the economic perspective, nuclear energy and coal are regarded as the most cost-effective energy sources. Their average generating cost is $0.025 /kWh, only half of the average generating cost for oil.

A nuclear power plant is one of the safest workplaces with less than one accident per 200,000 work-hours, which is lower than that for the finance, insurance or real estate industries, and far lower than that for the manufacturing industry, where the average is 45 accidents per 200,000 work-hours. As nuclear energy produces less CO_2 and other harmful materials than other energy sources do, its additional value is much higher than other energy sources. See Table 2 in Chapter 1.

Finally, as is shown in Table 1 in Chapter 1, from the perspective of life and health, only 0.4 day will be taken off one's expected life even if a person resides near a nuclear power plant all his/her life, which is much lower than the average 435 days taken off one's expected life by other accidents. No matter from which perspective, if you make a careful evaluation of reliability, nuclear energy is undoubtedly the most economical, space-saving and cost-effective energy resource on earth. We should attach importance to the nuclear energy.

In mid-1990s, a heated discussion was held in Idaho in the US over a plan to increase nuclear reactors so as to promote the construction industry there. However, the people of Idaho rose against the plan. It was interesting to learn that they were not necessarily against the after-effects of the nuclear energy industry, but against man-made pollution brought about by the popularization of nuclear energy.

Why are there so many anti-nuclear activists? Nuclear is a very sensitive word. When MRI (Magnetic Resonance Imaging) was invented, people were wise enough not to adopt the term "Nuclear Resonance Instrument". Otherwise this modern medical instrument, which has benefited a lot of people, would be doomed.

Conclusion

In the discussion above, I mentioned that Yan Hui was a most outstanding scholar who made the least mistakes in the world. His unreliability is 1%. He is much more reliable than other human beings.

Roughly speaking, the probability of causing an accident in driving is 1/10,000. Its reliability is much higher than Yan Hui and all the other human beings. In Iraq in 2009, the mortality rate of US soldiers was 1/10,000. The total mortality rate in all battles in the 20th century is about 1/100,000. During the Boston Marathon held on April 15, 2013, two pressure cooker bombs exploded, killing 3 people. The mortality rate of the marathon runners is generally in the order of 1/100,000. A Cola Company announced at a press conference following the so-called Cola incident in 1993 when a syringe needle was found in a coke can that the probability of finding foreign matter in such a can was 1/1,000,000. It was a serious mistake and caused a great uproar. How could a syringe needle get in a coke can? The accident rate for a computer system and software is 1/100,000. And nuclear facilities? It is one out of tens of millions.

Of all the daily operative systems, the most unsafe is the human being, which has a profound effect on society. Some people are against nuclear energy, but they forget that if a nuclear energy facility is well maintained, its accident rate is the lowest when compared with other daily life-related factors. Human beings make mistakes every day. Why don't we have an in-depth review of this phenomenon? If we improve ourselves and act 1/10,000 as well as Yan Hui, all the systems will be much safer. In a very big system we must find out where the bottleneck is and conduct radical reforms there.

Of all the systems, the human being is the most unreliable factor. The human being is the biggest bottleneck in enhancing our living quality. With the improvement of the human factor, much better results will be yielded.

I would like to quote Confucius as my conclusion. Confucius said, "A benevolent person will live a long life and a wise person will live a happy life." Does a long life mean happiness? The reliability of a rocket hitting its target is only 25%. Even if it hits an enemy warship, it will not necessarily sink the warship.

In other words, reliability has two implications.

The first implication is whether a life can be prolonged. That's what this appendix is about. The other implication is what efficacy a prolonged life will have. In the quotation "a benevolent person will live a long life", a long life means reliability. In the quotation "a wise person will live a happy life", a happy life means efficacy or the happiness we enjoy. I hope our society will emphasize the role of reliability in the evaluation of our life.

The twofold purpose of the study of reliability is to prolong a lifespan on the one hand and to improve its efficacy on the other. We must create a healthy and comfortable environment for the welfare of the populace. Don't let fruitless arguments threaten our well-being as well as our lofty aims.

Therefore, our ancient saint's teaching — "enslaving rather than being enslaved in material applications" – refers to the acme of reliability.

Appendix II

I made a speech at the Hong Kong Book Fair on July 23, 2011, which was followed by a dialogue between me and Hong Kong's noted journalist Chip Tsao who acted as the interviewer. The following are excerpts from my speech and my dialogue with Chip Tsao and they were carried in the *AM730* newspaper on August 3.

Analyze the Crisis and Opportunity in the Aftermath of the Fukushima Nuclear Accident

Japan's nuclear accident on March 11, 2011 has given people much cause for reflection on the safety of nuclear power. Way Kuo, President of City University of Hong Kong and a world-famous expert on safety and reliability, holds that crises give an impetus to the steady advance of humankind. Scientists are studying how to turn hydrogen into a common energy resource since it is the most environmentally friendly energy. Scientists are also carrying out vigorous research into a new-type of nuclear power plant which can completely consume the nuclear fuel.

Way Kuo illustrated the new development of energy sources and nuclear energy in his speech titled "Starting from a story about the man from the Qi Kingdom who feared that the sky might fall to an in-depth analysis of the 'crisis' and 'opportunity' of the Fukushima nuclear accident" to over 200 attendees at the recent Hong Kong Book Fair. He said, "The hydrogen which caused explosions in Japan's nuclear power plant is actually the cleanest energy, and has reached the 'acme of perfection'. Scientists are studying the possibility of turning hydrogen into a common energy resource. In dozens of years' time people may bring home a tank of hydrogen and use it to cook meals."

The treatment of nuclear fuel is a thorny problem facing every nuclear power plant. Way Kuo said a traditional nuclear power plant only consumes several percentages of the nuclear fuel. The remaining 90% or more of the unconsumed nuclear fuel is most dangerous. Scientists are working vigorously to design a new-type of nuclear power reactor which can completely burn up the nuclear fuel.

He cited an everyday example to illustrate this point, saying, "When you have a barbecue next time, do remember to use the right amount of charcoal for the meat you intend to eat. Don't put in more charcoal than you need. The complete burning up of the charcoal will save on water on the one hand and prevent fire on the other." He pointed out that it's important to select the appropriate site for a nuclear power plant, and make sure it is not built in a densely populated area.

For a time after the Fukushima tragedy, the Hong Kong Observatory detected radioactive substances in the air, which shocked Hong Kong residents. Way Kuo said normally a city gives off 0.05 units of radiation per hour (i.e., micro Sv/hr), while the background radiation in Hong Kong is usually 0.25 units. After the nuclear accident in Japan, the background radiation in Tokyo reached as high as 1.0 unit.

Way Kuo explained that the background radiation found in Hong Kong is fairly high because the dense high-rises contain a large amount of radioactive radon. In addition, the cinders used in reclaiming land from the sea may release highly radioactive substances. When asked whether the reclaimed land emits radiation, he answered that it depends on the thickness of the cement pavement.

After the nuclear accident in Japan many people proposed using or developing alternative energies. Way Kuo noted that so far nuclear power and hydroelectric power are the cleanest energies.

In order to implement the principle of safety and environmental protection, coal should be given up first of all. Statistics reveal that in 2010 alone 100,000 to 200,000 people died from coal-related accidents. As for solar-power generation, he noted that solar wafers are toxic and polluting.

Though scientists are studying how to develop non-polluting wafers by biological means, they have not yet achieved success. Wind-power generation will also produce CO_2. "For the time being, nuclear energy is known as an environmentally friendly energy. Will it be environmentally friendly in the future? I have no idea. It may change in future."

There are more than 400 reactors in the world. Nuclear power accounts for 20% of the electricity consumed in the world. In Hong Kong 25% of electric power comes from nuclear power. Nuclear power accounts for 80%, 30% and 20% of the electricity consumed respectively in France, pre-Fukushima-accident Japan and the US. If the nuclear accident had not occurred, Japan would have increased the ratio to 50%. The radiation leakage from the Fukushima nuclear plant made some countries give up or put off developing nuclear power. Germany has announced it would shut down all its nuclear power plants by 2022. China and Japan have decided to delay developing nuclear power.

Way Kuo agreed that nuclear power is by no means 100% safe. People can refuse to use nuclear power, but they still haven't found a satisfactory substitute. "This is a matter of a trade-off. If people are against the use of nuclear power, we may agree with them. However, when a special measure was adopted in some places to restrict the use of electric power in winter, there were rising public protests."

The safety of nuclear power falls into two broad categories. Firstly, reliability means the probability of accidents occurring in a nuclear power plant; the lower the probability, the higher the reliability. Secondly, safety means whether or not it is possible for people to flee from an accident in a nuclear power plant. Way Kuo said, "What people worry about most is peace of mind instead of reliability or safety." Even if the probability is low and the reliability is high, people still feel ill at ease. This is human nature.

"The most unreliable and dangerous factors are human beings. If we are able to improve the reliability of human beings, our environment will become very safe," Way Kuo said. The most serious nuclear power accidents in history were the Three Mile Island nuclear accident in the US in 1979, the Chernobyl nuclear accident in the former Soviet Union in 1986 and the Fukushima nuclear accident in Japan in 2011. These three accidents are related to human errors.

He said geographically speaking the nuclear power plant nearest to the earthquake epicenter in Japan is the Onagawa plant rather than the Fukushima plant. "It was decided that the 40-year-old Fukushima Daiichi Nuclear Power Plant was to be decommissioned by the end of March. Unfortunately, the nuclear accident happened before that could happen. However, the Onagawa nuclear plant had been in service for less than 30 years. Everything on earth will deteriorate and the problem of aging must be addressed. The Fukushima nuclear plant did not deal with this problem properly."

Way Kuo concluded his speech with a soul-stirring story: "Idaho is where the US built its first nuclear reactor after World War II. It's a sparsely populated and very beautiful place. When the US government planned to build a heavy-duty nuclear power plant there, the plan was turned down in referendum by the local people. When asked by a reporter why they were against the plan, they answered, 'We don't care about the building of a nuclear power plant. What we're worried about is that the over 1,000 workers whom the new nuclear power plant is going to employ will undoubtedly cause pollution and increase crime.'"

A Dialogue with Chip Tsao

After delivering his speech, Way Kuo discussed nuclear power and radiation with Chip Tsao and other participants. The following are excerpts from the dialogue. Hereafter Chip Tsao and Way Kuo are referred to respectively as Tsao and Kuo.

Enjoy a Hot Spring and Eat Seafood

Tsao: Hong Kong residents like to enjoy a hot spring in Japan's Hokkaido in winter. Is the sulfur spring radioactive? Before, our skin would become shiny and smooth after bathing in a hot spring. I wonder whether blisters will emerge on our skin now.

Kuo: A hot spring is radioactive. The deep-sea fish and shellfish that Hong Kong residents like eating may also be radioactive.

Tsao: Some experts said the water in the reactors flowed into the sea and reached the beach of Northern Philippines. As a result, several hundred sea turtles died. As the current has reached Taiwan and Hong Kong, can we eat the nearby seafood anymore?

Kuo: It's easier to detect radioactive substances than plasticizer. It takes several days to detect plasticizer in Taiwan whereas the detection of radioactive substances only takes one second. Scientifically speaking, our environment hasn't been affected.

Tsao: Plasticizer will make a certain organ of a boy grow especially small while radiation does not produce such a side effect. So, as far as pollution is concerned, I'd rather choose radiation.

Japanese "Nuclear Warriors" Didn't Die

Local resident: You illustrated reliability with the examples of a traffic accident, bridge collapse, deaths after surgery and an aircraft crash. I don't think they have anything to do with me. Fukushima nuclear crisis has killed more than 10,000 people. Can you share your views with us?

Kuo: First of all, I beg to differ. In my opinion, traffic accidents, bridge collapses, deaths after surgery and aircraft crashes are all closely related to everyone's life. I don't want to make further comment here about this.

The safety of nuclear power can be divided into three aspects: reliability, safety and peace of mind. People are most anxious about the "peace of mind".

After the big earthquake on March 11 in Japan, the international and local media reported that the 50 Japanese nuclear warriors would die within two weeks. As a matter of fact, so far not a single one of them has died. During the Three Mile Island nuclear accident in 1979, which was initially more serious than the Fukushima nuclear accident, nobody died, either. About 32 years have passed since then, and still no evidence shows that the people who lived within a radius of 50 km of the Three Mile Island Nuclear Power Plant are more likely to develop cancer. Nobody knows what will happen in 35 years, though.

iPhone 4 and Radiation from Reclaimed Land

Tsao: A report from the UK said that the radiation of iPhone 4 was damaging to the human brain. Is this true?

Kuo: The iPhone is not really radioactive. And the background radiation from the iPhone is different from that of Fukushima. The former is soft radiation while the latter is the strong cancer-causing radiation composed of α, β, γ rays. Compared with the nuclear power plant, soft radiation has not been proven to be harmful to health.

Tsao: What do you think of Guangdong's plan to build a dozen more nuclear power plants? Is Hong Kong's high background radiation from the nuclear power plants in Guangdong?

Kuo: The supply fails to meet the demand for energy sources in China. That's why the houses of western China are dimly lit. According to the data that I have collected, China is increasing the pace in developing nuclear power. In so doing we must balance the following three factors: reliability and safety, the energy source, and economy and environmental protection.

Nuclear power is by no means 100% safe. Nevertheless, both nuclear power and hydroelectric generation are considered the safest energies. Coal is the most unsafe energy. Last year as many as 200,000 people died in coal-related accidents.

Why is the background radiation of Hong Kong so high? Besides environmental pollution, the cinders used in reclaiming land from the sea release highly radioactive substances.

Tsao: If we drive piles into the bed of the sea, will they still emit radiation?

Kuo: It depends on the thickness of the cement. If it is thick enough, there won't be any radiation leakage. Otherwise, it still gives off radiation.

Residents Nearby a Uranium Mine Live a Longer Life

Tsao: Can the human body produce the antibody for radiation? When we drink distilled water at a hotel in India, we will have loose bowels for three days and nights. But the local people won't get sick even when they take a bath in the river.

Kuo: Your theory is similar to that of radiation biologists. As this is not in my line, I'm not in a position to offer my opinion. Strictly speaking, there is no consensus of opinion about the impact of low dose radiation on organisms. There is a uranium mine in Iran where nearby residents enjoy a longer life than the average Iranian. This is a fact. Does it have something to do with their genes? I'm not sure, because a thorough analysis must be performed. Hong Kong residents have a fairly long life. When I came to live in Hong Kong, I often joked about the possibility that I may live longer here.

Way Kuo is Realistic and Truthful

Local resident: You mentioned the actual safety and psychological safety of nuclear power. The actual safety is the concern of specialists and scientists. For us ordinary people, we think more of psychological safety. How should the government and the nuclear power plant act to make people feel safe psychologically?

Kuo: It's important to spread scientific knowledge.

Tsao: That's why we invited President Way Kuo to give a speech.
                  Just as he is called Way Kuo, President Kuo has a "way"
                  to convince you. Unlike us who "brag" and "bluff" on the
                  radio, Way Kuo is a scientist who speaks on the basis of
                  solid evidence.

Death of Luo Fu's Son

Tsao: Luo Haixing, the son of Luo Fu, the former editor-in-chief of *The
New Evening Post*, died of leukemia last year. I asked him whether
anybody of his clan had ever suffered from leukemia. He said, "None".
But he suspected that it had something to do with his imprisonment
in a jail in Guangdong after the June 4th incident. There was a uranium mine near the prison and its construction came to a standstill
before it was completed. He was held in captivity for more than a
year. When a new guard came to bring meals to him, he asked him
about the previous guard. The new guard told him that the previous guard had died of cancer. And several others died of the same
disease. I asked him whether there is any radiation leakage in the
uranium mine. He said he had no idea.

Postscript

On March 11, 2011, Odaka in Minamisoma, a district over 10 km from the Fukushima Daiichi Nuclear Power Plant, was hit by a triple blow — an earthquake followed by a tsunami and nuclear radiation leaks. Time in the village seemed to stand still in the immediate aftermath of the disasters.

The reconstruction efforts were slow. At the time of publication, rusty cars and debris of equipment washed away by the seawater still littered the countryside. Due to excessively high levels of radiation, the area within a 20 km radius of the Fukushima nuclear plant was a restricted area. More than 10,000 inhabitants in Odaka had to be relocated after the accident and lived a nomadic existence for a year. More than 100,000 refugees, who used to live near the plant, are still living in fear and anxiety, worrying about the safety of their homes. It will take two to three decades for the radiation from the plant to return to the normal level.

Don't we all have an opinion about nuclear energy in our pursuit of energy and environmental protection?

The Japanese proved themselves to be calm, orderly and law-abiding citizens even when faced with the painful triple blow of an earthquake followed by a tsunami and nuclear radiation leaks. People from different countries voluntarily reached out to help the victims, which are admirable. For different reasons, governments around the world responded to the disaster with a range of policy changes regarding nuclear power such as phasing it out, suspending it and continuing its use. Such mixed responses are thought-provoking. The public panic and overreaction in countries and regions neighboring Japan make us realize the importance of understanding nuclear energy and knowing the whole truth about energy needs.

So far, Germany, Switzerland, Belgium and Italy have decided to abandon nuclear power; US, China, South Korea, UK, France and Czech Republic have pledged to continue using nuclear power. Finland was the first country to announce that it will build a nuclear power plant after the Fukushima accident.

As early as July 2011, the UK announced plans to build eight new nuclear plants equipped with 10 or more reactors with a total investment of £50 billion (about $80 billion), and issued the first license to build a nuclear power plant in November 2012. Sweden, which adopted a phase-out policy following a referendum in the 1980s, decided nevertheless to extend the operation of its nuclear reactors. The nuclear power giant, France, also said it will continue to invest €1 billion (about $1.3 billion) in nuclear power.

In February 2012, the NRC in the US approved the construction and operating license for two additional units, Unit 3 and 4, at the Vogtle site in Georgia, which is expected to cost $14 billion. The approval marked the first construction license ever issued for a US nuclear plant since 1978, a year before the Three Mile Island accident. Meanwhile, there are other plans regarding 17 units in 11 plants that are waiting for the NRC's approval. South Korea is also developing its nuclear energy at full speed ahead and is exporting its nuclear power facilities, hoping to become a nuclear energy power of the 21st century.

Following the Fukushima accident, the Chinese government announced the suspension of approvals of new plants and ordered thorough safety checks on the operating nuclear reactors as well as those under construction. Then one year later, the Chairman of China's State Nuclear Power Technology Corporation predicted in a press conference held on March 10, 2012 that "the government will resume the examination and approval of nuclear power plant projects this year." He also pointed out that "China needs to produce more nuclear power to meet the needs of economic and social developments because of the shortage of primary energy, the strong market demand in energy and the imbalanced energy structure."

Four of mainland China's newly-built nuclear reactors are imported from the US and deployed with the most advanced technology (Generation III or III$^+$ reactors or the AP1000 technologies) and their safety standards are far higher than those currently set by China and the IAEA. When the fourth reactor is built, Generation III reactors and the rate of nuclear equipment and key materials produced at home will be over 80%. On the other hand, China is also developing its own technology for high temperature gas-cooled reactor (HTGR), the Generation IV reactors. One of the most important features of these reactors is its inherent safety, which can ensure a safe and reliable operation of the uranium-hydrogen reactor unit without any constraints and limits of protection conditions.

During the development of nuclear power, concerns about operation and security have increased. We have paid a heavy price in Fukushima accident, and it has alerted those working in the industry to be highly cautious and be ready to stand by to prevent any extreme events from happening.

In China, there is still great room for improvement in safety regulations concerning nuclear power and the transparency of information release. There is the need for a clearly defined notion of putting safety first in nuclear power expansion, and making real efforts to strengthen emergency preparedness against any possible major nuclear accidents with a scientific mindset and method.

The seasons come and go. More than two years have gone by since the Fukushima tragedy in March 2011. This book will hopefully serve as a reminder of the need to learn about the development of energy and environmental protection. The excerpts below are translations of three important news stories. They may prove useful when we reflect on these questions.

Bill Gates Discussing New Nuclear Reactor with China

Microsoft co-founder Bill Gates was in discussions with China National Nuclear Corporation (CNNC) to jointly develop a low-cost and safer new type of nuclear reactor which generates very little waste, according to The Wall Street Journal on December 8, 2011.

The news reported that Gates said he had funded a company, TerraPower, which has been in talks with CNNC about setting up a traveling wave reactor (TWR) since 2009.

TWR is a new class of nuclear reactor that can run on depleted uranium. When it is built, TWR will be a nuclear reactor that is smaller in magnitude and less-polluted and can run without adding fuel for many years. TerraPower has been looking for a country that is willing to build the first local TWR. Based on the principle of fast-neutron, the concept of TWR has been known for years, and TWR can directly convert depleted uranium into usable fuel with a utilization rate of 60% to 70%, and produce significantly smaller amounts of nuclear waste than conventional reactors.

Since the Fukushima accident, China has been committed to addressing the challenge of the safety risks in nuclear power and the new nuclear reactors are deemed as the hope and key to the future direction of nuclear power expansion.

Since retiring from Microsoft Corp., Gates has actively promoted research and development of relevant technologies of clean energy. One of his three dreams, it is said, is to provide affordable and clean energy for the poor, and he also stated that nuclear power is the one that has the most potential in the energy field.

Technical Lessons Learned from the Fukushima Catastrophe

Eight months after the Fukushima nuclear accident, the *IEEE Spectrum* published a special report titled Fukushima and the Future of Nuclear Power by Eliza Strickland in November 2011. The report gave a blow-by-blow account of the 24 hours at Fukushima where the worst nuclear accident since Chernobyl took place, and just hoped people could learn lessons from this catastrophe and find out the hidden dangers, and in the end improve the safety of nuclear plants.

The author's report is "based on interviews with officials from the Tokyo Electric Power Co. (TEPCO), Japan's Nuclear and Industrial Safety Agency, the US Nuclear Regulatory Commission, the International Atomic Energy Agency, local governments, and with other experts in nuclear engineering, as well as a review of hundreds of pages of official reports."

The report said, "…the chain of failures that led to the disaster at Fukushima was caused by an extreme event. It was precisely the kind of occurrence that nuclear-plant designers strive to anticipate in their blue-prints and emergency-response officials try to envision in their plans. And in the end the calamity will undoubtedly improve nuclear plant design."

Eliza Strickland identified six main lessons about the Fukushima nuclear accident after giving a description of a series of failures. In a word, she said, the emergency cooling water and exhaust system should have been improved. We can see from these lessons that sometimes making a few simple changes in procedures and operations can significantly improve the safety of nuclear power. If the workers had been able to vent the gases in the reactor sooner, the destruction of the rest of the plant might well have been averted.

Japan's NAIIC Report

In early July 2012, the 13-member Japanese National Diet of Japan Fukushima Nuclear Accident Independent Investigation Commission (NAIIC) chaired by Kiyoshi Kurokawa, a professor from University of Tokyo, issued a detailed report of 600-plus pages on the Fukushima nuclear accident after a six-month investigation. The report criticized the collusion between government officials and business, and also disclosed the Nuclear Industrial Safety Agency (NISA), as part of the METI of the government, "covered up" the nuclear safety issues; also the TEPCO management deliberately hid the potential risks of nuclear power from the public and requested the government to diminish the prospect of unprecedented events such as the tsunami would hit the areas around Fukushima.

The report concluded that the Fukushima accident was "a profoundly manmade disaster" and it was "made in Japan." It also pointed out, "fundamental causes (of the accident) are to be found in the ingrained conventions of Japanese culture: our reflexive obedience; our reluctance to question authority; our devotion to 'sticking with the program'; our collectivism; and our insularity."

It is no wonder that such conclusions were drawn in the report. They perfectly match the comments and analyses I made on the Japanese society in the first edition of the book which was published on the first anniversary of that tragic event.

My Reflections

We should insist on the transparency and openness of information whether the issues at hand are related to industrial safety, environmental protection or energy security. Just as I observed shortly after the Fukushima accident, there was one common factor in the tragedies in Fukushima and Chernobyl of former Soviet Union: the companies and the related government regulators withheld key information from the public. Because of the incomplete information they had at their disposal, the panic-stricken public soon began to see all nuclear power issues in the light of the horrors and nightmares of radiation. To make matters worse, the operator of Fukushima nuclear plant proceeded to issue to the Japanese and the global public frustratingly vague information.

The massive tsunami triggered by the earthquake in Japanese waters in 2011 and the consequential nuclear accident were unanticipated in *WASH1400* or *SOARCA*, the authoritative analyses on nuclear safety. The Fukushima nuclear accident will undoubtedly offer an opportunity for mankind and the nuclear industry "to learn from mistakes", and will boost the continuous development of energy science.

We have to solve real problems before we can move forward. The issue of energy involves not just science and technology but also environmental protection, economics, politics and social psychology. Take Germany's announcement of abandoning the use of nuclear energy, for example. A January 2012 study found that 20% of German companies had started to shift their business abroad due to the worry about the possible soaring prices of raw material caused by the government measures. Such a scenario was unexpected.

I will quote Zhao Gu's poem "Hearing the flute" as the epilogue of my visit to Sendai.

> *Who's playing the flute yonder*
> *In that painted tower, I wonder?*
> *Loud and soft the tune's coming near,*
> *As on and off the wind's blowing here;*
> *It soars high to the heavens, where the clouds no longer sail,*
> *And lands cool on the curtains, where the Moon shines pale.*
>
> *Then, the legendary Huan the Flutist springs to mind,*
> *And Ma the Poet, whose "Ode to the Flute" is the best I find.*
> *Now the melody's over, who knows if the Muse is still there,*
> *While the song from that instrument lingers in the air?*

With their robust finances and expensive real estate, Hong Kong and Shanghai have a great demand for energy and set high standards for its safe use. The 21st century is a time of scientific and technological civilization. As a commercial mega-city, Hong Kong, the Pearl of the Orient, boasts first-rate financial and monetary professionals, but does not attach enough importance to scientific and engineering expertise.

The Fukushima nuclear accident demonstrates the close relationship between science and engineering on the one hand, and people's daily life and well-being on the other. Hong Kong is no exception. The current situation has taught us a good lesson, i.e., if people at large are worried about safety, we must seek scientific explanations and let the evidence speak for itself. This is the only way we can avoid becoming unnecessarily jittery.

Here I would like to offer some of my random thoughts:

1. Many chapters in this book were written before the collision between the two high-speed trains in Wenzhou, China, on July 23, 2011. When Japan's Fukushima nuclear accident occurred in March, I pointed out that China's railway authorities should look into the safety of the trains because these high-speed trains had developed at an inconceivable speed in China. Industrial safety and quality control are not usually the focal points in mainland China's culture. If due

attention is not paid to the importance of reliability, we will witness more tragedies of this kind.

2. Hong Kong, the city that never sleeps, is responsible for severe light pollution. About 20% of cheap electricity consumed in Hong Kong comes from the mainland's nuclear power plant. How do people in Hong Kong reconcile their huge consumption of electricity with their dismissal of the contribution of nuclear energy in generating electricity? In order to have a comprehensive knowledge of energy sources, we must give equal emphasis to energy conservation and scientific innovation. The prosperity of finance and real estate market relies on safe and reliable energy sources.

3. There are 20 nuclear power plants in Guangdong but no programs are devoted to the study of nuclear energy and safety in either Hong Kong or Shenzhen. There is therefore a need for City University of Hong Kong, a university focused on professional education, to serve society by providing basic information on nuclear energy and nuclear power safety, and to undertake teaching and research activities on the assessment and management of nuclear energy. On February 15, 2011, the University proposed to the University Grants Committee (UGC) to establish a program for nuclear power and risk engineering.

   At the time, some people at the University were extremely doubtful that the government would approve of the proposal. To their surprise, after the Fukushima nuclear accident on March 11, the UGC approved the University's plans. Society tries to scrimp on innovation sometimes. Safety and reliability are not usually people's primary concerns when there is no immediate profit to be made. However, rather than waiting for people to come to our aid, it is better to depend on ourselves. Energy sources, nuclear safety and risk engineering are a reality that every advanced society must face. People in Hong Kong should study and research these special fields.

4. Mainland China is suffering from power shortages while Taiwan and Hong Kong, without a comprehensive policy on energy, environment protection, science and technology, waste huge amounts of energy. Industrial safety accidents occur frequently in the cross-Strait areas where there is a lack of understanding of the concepts of safety and reliability. In this world where food waste and global food shortages exist side by side, our society has yet to come up with

a solution that achieves an ideal balance of humanity's welfare, economic development, environmental protection and industrial safety.

5. While we develop nuclear power, we should also invest in the research and development of renewable energy sources. According to China's National Energy Administration, the solar energy exported from China accounts for over 50% of the world trade volume of this energy, and China became the top solar energy producer in the world in 2012. In the meantime, China's total installed capacity of wind power reached 65 gW in 2010, ranking the first in the world and becoming a dominant manufacturer of wind power equipment. That's why the impacts of the energy of the wind and the sun on the environment cannot be ignored.

Energy deficiency is a pressing issue, one that confronts all human societies. The whole spectrum of energies — hydropower, thermal (coal, oil & natural gas), nuclear, wind, solar, biofuel and others (geothermal, ocean energy and marsh gas)— have their strengths and weakness in terms of efficiency, safety, reliability, environmental protection, reserves, and economic value. Some are legitimate concerns, while others are unjustifiable fears, and even exploited and manipulated by politicians.

A Story about Ted Kennedy

In 1980, US Senator Edward Kennedy of Massachusetts, a younger brother of John F. Kennedy, the President of the US who was assassinated in office, announced his candidacy for the American presidency, challenging incumbent President Jimmy Carter for the Democratic nomination for the 1980 election.

As it was not long after the Three Mile Island accident, he took advantage of it by making anti-nuclear energy and power one of his main campaign themes. Regrettably, an incident from Ted Kennedy's past came back to public attention. On July 18, 1969, Ted Kennedy drove his car off a bridge into a tidal channel on Chappaquiddick Island, Massachusetts. He got out of the vehicle but left behind the woman who was in the car with him. His behavior provoked public outrage.

In reaction to his anti-nuclear stance, there appeared in the Los Angeles International Airport a banner which referred scathingly to the Chappaquiddick Incident, pointing out that fewer people died from nuclear power than in Ted Kennedy's car.

Rather than just a mockery of Ted Kennedy, the banner was accurate with the numbers in 1980, a year after the Three Mile Island. It is still true 33 years later in 2013 in the US.

In evaluating energy, we must adhere to strict principles, taking into consideration environmental protection, life-cycle pollution, reliability of energy, and the risks that people have to face. All these issues were highlighted by the Fukushima nuclear accident. To tackle them, we need to pull together the insights of various disciplines, work across multiple domains, and seek to gain a clearer picture of the relationships between the development of science and technology, sustainability and social well-being.

The Fukushima nuclear accident has reminded us again that history is written and illuminated by crises. As far as energy consideration and scientific innovation are concerned, there is a long way to go. Besides developing new energy sources, conserving energy should be one of our top priorities.

A Trip to Fukushima on July 2, 2013

On July 2, 1968, I walked out of the red-brick building of Taipei Municipal Jianguo High School, and walked through Taipei Botanical Garden in high summer. Withstanding the onslaught of pollen against my sinusitis, I stepped into the examination hall which had no air-conditioners or ceiling fans. Drenched in perspiration, I took my college entrance examination. Fortunately, I succeeded in entering the narrow gate of the Department of Nuclear Engineering of National Tsing-Hua University.

On the same date 45 years later—July 2, 2013—or more than two years after the Fukushima nuclear accident, I set foot on the soil of Fukushima upon invitation. It was another rainy day. Although Fukushima is different from the Jiangnan area of China, the scene that day reminded me of what Zu Yong wrote in his poem "Traveling South of the Yangtze", which triggered a chain of thoughts about my study tour of the post-disaster Fukushima nuclear plant:

> *The mountains of Chu extend endlessly,*
> *The only road home is bleak.*
> *By the sea, rain appears in the clear blue sky,*
> *The tides came in, roaring all night.*
> *My sword remains by the Southern Dipper,*
> *My letters buffet the distant North Wind.*
> *I would send gifts of Changsha oranges,*
> *But who will take them to the Luoyang Bridge?*

(Adapted from the *Poems of the Masters, China's Classic Anthology of Tang and Song Dynasty Verse.*)
http://www.mountainsongs.net/poem.php

The town of Minamisoma was still listed on July 2, 2013 as a Grade-1 disaster area with the average background radiation level between 0.7 and 5 mSv/hr (i.e., about 14 to 100 times the normal background radiation level). There were no pedestrians in the streets except some nuclear power plant employees or municipal maintenance workers. With the buildings, vehicles, bridges and factories which had been devastated by the tsunami, the place looked like a war-torn zone. Only those buildings protected by hillocks or dense forests survived the disaster.

> *The river in drizzles, its banks covered with grass so green,*
> *The Six Dynasties gone, no bird's cry would again win.*
> *Willows unfeeling stand on guard for all portals,*
> *They loom like mist adorning the Capital.*

(Translation by Jiang Shaolun)

Wei Zhuang's poem "The Ancient Capital" captured the feelings I had as I stood in front of the Fukushima Daiichi nuclear plant. Looking as far as my eyes could see, only the nuclear power plant stood alone in the lingering light of the setting sun. It was a spectacle of devastation. On August 10, TEPCO announced that contaminated groundwater in the plant had risen to 60 cm above the protective barriers and was now leaking into the Pacific Ocean.

With the railway in Minamisoma out of service for two years, the land along the tracks was overgrown with weeds. Wild flowers were in full bloom. The forests and lush green grass seemed to vie with each other for beauty. My trip to Fukushima was accompanied by light drizzle. Weeds grew around the nuclear plant. The streets were deserted, with only the sound of birds in the background. Is it true that the most unfeeling and heartless were the evergreen and unyielding pines and cypresses which stood by the side of Fukushima nuclear plant?

The Current Status of the Disaster-stricken Fukushima and its Prospects

The devastating earthquake on March 11, the unprecedented tsunami and the Fukushima nuclear accident haunted the patients who were hospitalized for treatment in the local Fukushima hospital. Dozens of inmates in their 70s and 80s were evacuated in emergency amid the misfortunes caused by the March 11 incident. However, poor communications, delayed transport, the long and difficult journey, and a severe scarcity of human support combined to add to the burden of these senior citizens and hasten their death.

The family dependents of the deceased have filed a charge in court against TEPCO and demanded compensation. This may be another chapter of the after-effects of the devastating earthquake. Not even the lawyers could have foreseen such a development!

Although WHO believed that the nuclear leakage would increase the risk of development of cancer among inhabitants residing in the area, UNSCEAR differed in its opinion.

After two years of extensive observations and careful analyses, dozens of UNSCEAR experts came to the conclusion that compared with normal background radiation, the residents of Fukushima were exposed to a fairly low dose of radiation which does not constitute a threat to their health. They even predicted that as proper protective measures had been adopted for those staff members who were exposed to high doses of radiation, it is unlikely that they would contract life-threatening or radiation-related acute diseases in the future.

If the prediction of the UNSCEAR experts proves to be right, the consequences of the Fukushima nuclear accident will be similar to those of the Three Mile Island incident that happened in 1979 in the US. No obvious casualties were recorded many years after the Three Mile Island incident.

Nuclear incidents often cause a rippling effect. In view of the Fukushima nuclear accident, the executive committee of the European Union (EU) put forward on June 13, 2013 a nuclear safety plan to enforce mandatory intensive inspections of nuclear plants. It demands that a legally binding review be carried out every six years and a safety inspection be taken at least every 10 years of all the 132 nuclear reactors in the territory of EU. The report, entitled "EU Nuclear Stress Tests: Legally Binding Reviews Every Six Years", is identical to the gist of this book: emphasis should be given to the importance of transparency in nuclear power operation.

The "stress tests" mentioned in the report is in accord with the proposal I made to Taiwan's No. 4 Nuclear Power Plant in Lungmen, which can be found in the second edition of this book. If safety regulations are strictly enforced, there will not be any nuclear radiation leakage.

Opinions differ widely about the impact of the Fukushima nuclear accident on human life. However, it is an undisputed fact that not a single person has died of radiation during the past two years since the Fukushima nuclear accident. There is a world of difference between this result and the reports of the media or the hearsay on the Internet.

In my opinion, we should agree to disagree and wait to see the forthcoming results. Cheng Hao put it best in his poem "An Impromptu Poem Composed in Fall":

> *A man indifferent to vanity does everything in an unhurried manner,*
> *Waking up in the morning when the sun shines through the east-facing window.*
> *Observed in tranquility of mind, everything runs their natural course,*
> *What governs the four seasons governs us all.*
> *The same principle applies to heaven, earth, and the universe,*
> *This comprehension helps one see through tumultuous winds and storms.*
> *He who can endure hardship and resist the seduction of sensual pleasure,*
> *Is truly worth to be honored as a wise man.*

> (Translation adapted from LC Wang Press at: http://www.lcwang-press.com/poems-song/impromptu-poem-written-in-autumn.html)

When it comes to our own safety and well-being, let us be guarded in what we say. If you want to know how the story will unfold, please wait to listen to the analysis of the experts.

Additional Reading Materials

1. *Blue Book of Cities in China: Annual Report on Urban Development of China*, No. 5, Beijing: Social Sciences Academic Press, 2012
2. B. L. Cohen, *Before It's Too Late: A Scientist's Case for Nuclear Energy*, New York: Plenum Press, 1983
3. The Royal Society Science Policy Centre Report 10/11, *Fuel Cycle Stewardship in A Nuclear Renaissance*, London: The Royal Society, 2011
4. D. Helm, The Carbon Crunch: *How We're Getting Climate Change Wrong - and How to Fix it*, New Haven: Yale University Press, 2012
5. F. H. Koch, *Hydropower-Internalized Costs and Externalized Benefits*, International Energy Agency (IEA), Ottawa, Canada, 2000
6. Way Kuo and X. Zhu, *Importance Measures in Reliability, Risk, and Optimization: Principles and Applications*, New York: Wiley, 2012
7. B. Le Patner, *Too Big to Fall: America's Failing Infrastructure and the Way Forward*, New York: Foster Pub., 2010
8. Pan Ziqiang *et al.* (ed), *Social Impacts of Nuclear and Radiation Accidents*, Beijing: China Atomic Pub., 2011
9. UNSCEAR 2008 Reports, *Sources and effects of ionizing radiation*, Vienna: United Nations Scientific Committee on the Effects of Atomic Radiation (UNSCEAR), 2008
10. *World Energy Outlook,* Paris: International Energy Agency, 2013
11. Q. Schiermeier, "An Ambitious Plan to Slash Greenhouse-gas Emissions Must Clear Some High Technical and Economic Hurdles," *Nature*, 156–158, Vol. 496, April 2013

Afterword

Let the chips fall where they may?
While some people were concentrating on their work,
Some were adhering to the principle of self-preservation,
Some were indulging in loud and empty talk,
And some others were even intimidating people by resorting to populist doctrine.
There are no more willow catkins rising with the breeze;
But only sunflowers that turn their heads toward the sun.
As people living in a free and democratic environment,
We should not let the chips fall where they may or follow others blindly.
We should not put aside industrial safety and innovation
Or simply talk big about energy and environmental protection.

Index